Electronic Media and Broadcasting

Electronic Media and Broadcasting

Edited by
Hubert Reeve

Larsen & Keller
www.larsen-keller.com

Electronic Media and Broadcasting
Edited by Hubert Reeve
ISBN: 978-1-63549-101-2 (Hardback)

Larsen & Keller
Published by Larsen and Keller Education,
5 Penn Plaza,
19th Floor,
New York, NY 10001, USA

Cataloging-in-Publication Data

Electronic media and broadcasting / edited by Hubert Reeve.
p. cm.
Includes bibliographical references and index.
ISBN 978-1-63549-101-2
1. Broadcasting. 2. Digital media. 3. Broadcasting--Technological innovations. 4. Mass media.
I. Reeve, Hubert.
HE8689.4 .E44 2017
384.54--dc23

Printed and bound in China.

Table of Contents

	Preface	**VII**
Chapter 1	**Introduction to Electronic Media and Broadcasting**	**1**
	• Electronic Media	1
	• Broadcasting	3
Chapter 2	**Broadcast Engineering: An Overview**	**10**
Chapter 3	**Evolution of Broadcasting**	**16**
	• History of Broadcasting	16
Chapter 4	**Fundamental Concepts of Broadcasting**	**38**
	• Radio Broadcasting	38
	• Digital Audio Broadcasting	45
	• Commercial Broadcasting	57
	• Public Broadcasting	59
	• Remote Broadcast	95
	• Microbroadcasting	96
	• Broadcast syndication	97
	• Broadcast Programming	113
	• Rerun	117
	• Simulcast	121
	• Broadcast Signal Intrusion	125
Chapter 5	**Applications of Electronic Media**	**132**
	• Journalism	132
	• Broadcast Journalism	139
	• Digital Journalism	142
	• Electronic Publishing	150
	• Mobile Content	153
	• Podcast	157
Chapter 6	**Advertising in Electronic Media**	**165**
	• Television Advertisement	165
	• Radio Advertisement	182
	• Mobile Advertising	186
	• Online Advertising	190
Chapter 7	**Production and Broadcasting**	**205**
	• Broadcast Network	205
	• Broadcast Automation	207
	• Outside Broadcasting	211

- Electronic Field Production 213
- Electronic News-gathering 216
- Broadcast-safe 221
- Broadcast Delay 225
- Bleep Censor 227

Permissions

Index

Preface

This book provides detailed explanation of the various concepts and applications of electronic media and broadcasting. It picks up individual branches and explains their need and contribution in the context of the growth of the field of journalism. Electronic media refers to any media that is spread through electromagnetic channels and broadcasting refers to the act of spreading this information using the electronic devices. In this book, constant effort has been made to make the understanding of the difficult concepts of electronic media and broadcasting as easy and informative as possible for the readers. Most of the topics introduced in this book cover new techniques and the applications of this subject. This book attempts to assist those with a goal of delving into the field of electronic media and broadcasting.

A short introduction to every chapter is written below to provide an overview of the content of the book:

Chapter 1 - Any media that uses electronics in the process of communication is known as electronic media and electronic media as such can be divided in analogous media and digital media. This chapter introduces the reader to electronic media and broadcasting. The chapter provides a comprehensive overview of electronic media and its types with examples to aid in a deeper understanding of the subject matter; **Chapter 2** - Broadcast engineering deals exclusively with radio and television broadcasting. The chapter provides a detailed summary about broadcast engineering, its importance, functions and its scope. There is also a section about the skills and knowledge that broadcast engineers are generally required to have. Electronic media and broadcasting is best understood in confluence with the major topics listed in the following chapter; **Chapter 3** - Broadcasting refers to the transmission of audio or video to an audience over electronic mass-media devices like radio, television and computers. This chapter chronicles the history of mass-media broadcasting and informs the reader about the progress that has been till date in the field. The reader will gain a deep understanding of this field; **Chapter 4** - The chapter strategically encompasses and incorporates the major components and key concepts of broadcasting, providing a complete understanding. Some of the topics discussed in this chapter include radio broadcasting, digital audio broadcasting, commercial broadcasting, public broadcasting, remote broadcast, microbroadcasting, broadcast syndication, simulcast, broadcast signal intrusion etc.; **Chapter 5** - Electronic media is used in the fields of digital journalism, broadcast journalism, electronic publishing, mobile content and podcasts. This chapter discusses these applications and the impact of electronic media in revolutionizing the field of communication for the publisher as well as the reader. The topics discussed in the chapter are of great importance to broaden the existing knowledge on electronic media; **Chapter 6** - Electronic media has several advantages over traditional print media and advertising has exploited this. One is bombarded with advertisements on the radio, television, mobile and internet. This chapter focuses on this aspect and educates the reader about the format, design, advantages, disadvantages and techniques used to advertise over electronic media; **Chapter 7** - Production is an integral part of broadcasting and this chapter discusses production terminology like broadcast network, broadcast automation, outside broadcasting, electronic field production, electronic news-gathering, broadcast-safe, broadcast delay and bleep censor. The reader is informed about the production side of broadcasting and its importance.

I extend my sincere thanks to the publisher for considering me worthy of this task. Finally, I thank my family for being a source of support and help.

Editor

Introduction to Electronic Media and Broadcasting

Any media that uses electronics in the process of communication is known as electronic media and electronic media as such can be divided in analogous media and digital media. This chapter introduces the reader to electronic media and broadcasting. The chapter provides a comprehensive overview of electronic media and its types with examples to aid in a deeper understanding of the subject matter.

Electronic Media

Electronic media are media that use electronics or electromechanical energy for the end user (audience) to access the content. This is in contrast to static media (mainly print media), which today are most often created electronically, but do not require electronics to be accessed by the end user in the printed form. The primary electronic media sources familiar to the general public are video recordings, audio recordings, multimedia presentations, slide presentations, CD-ROM and online content. Most new media are in the form of digital media. However, electronic media may be in either analogue electronics data or digital electronic data format.

A screenshot of a web page. The computers to store, transmit, and display the web page are electronic media. The web page is an electronic medium.

Although the term is usually associated with content recorded on a storage medium, recordings are not required for live broadcasting and online networking.

Any equipment used in the electronic communication process (e.g. television, radio, telephone, desktop computer, game console, handheld device) may also be considered electronic media.

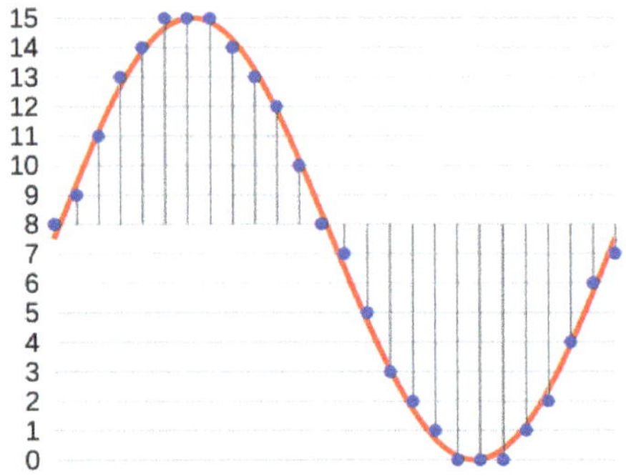

Graphical representations of electrical audio data. Electronic media uses either analog (red) or digital (blue) signal processing.

History of Development

- **Transmission**
  - ***Wire and transmission lines***
    - ☐ Telegraph 1795–1832
    - ☐ Facsimile 1843–1861
    - ☐ Telephone 1849–1877
    - ☐ Coaxial cable 1880
    - ☐ Fiber Optics 1956–1970
  - ***Wireless***
    - ☐ Radio 1897–1920
    - ☐ Satellite 1958–1972
    - ☐ Free Space Optics 1960s
  - ***Internet***
    - ☐ Downloading 1969 (first protocols for transferring files)
    - ☐ Live Streaming 1996 (RTP protocol)
- **Display and output**
  - Information Processing 1940s (Term)
  - Galvanometer 18
  - Telegraph Sounder 1844
  - Telephone Receiver 1849–1877
  - Light red 1801–1883
  - Neon 1893–1902
  - Teleprinter 1910
  - CRT 1922
  - Radio/television tuner 1894–1927
  - Speaker/Headphones 1876–1928/1930s
  - LED/LCD 1955–1962/1968
  - Laser light show 1970s
  - Computer Monitor 1950s/1976 (for PCs)
  - Large Electronic Display 1985
  - HDTV 1936 (Term) 1990s (Standards)
  - HMD 1968–current
- **Electrical Signal Processing**
  - Capture 1745 (Capacitor)
  - Analog methods of Encoding 1830s (Morse code)
  - Electronic Modulating 1832–1927
  - Electronic Multiplexing 1853 (TDM)
  - Digitizing 1903 (PCM telephone)
  - Electronic encryption 1935–1945
  - Online routing 1969
  - Electronic programming 1943–current
- **Electronic information storage**
  - ***Recording medium***
    - ☐ Punched card and paper tape 1725/1846
    - ☐ Phonograph cylinder and disk 1857–1958
    - ☐ Film 1876–1889
    - ☐ Magnetic storage 1898–2003
    - ☐ RAM 1941–current
    - ☐ Barcodes 1952/1973 (UPC)
    - ☐ Laser Disc 1969–1978
    - ☐ Compact Disc/DVD 1982/1993–current
  - ***Content formats***
    - ☐ Content (media) 1877–current
    - ☐ Audio recording 1877–current
    - ☐ Video recording 1952–current
    - ☐ Digital file formats
    - ☐ Database content and formats 1963–current
- **Interactivity**
  - Control panel
  - Input device
  - Game controller
  - Handheld
  - Wired glove
  - Brain–computer interface (BCI)

Uses

Electronic media are ubiquitous in most of the developed world. Electronic media devices have found their way into all parts of modern life. The term is relevant to media ecology for studying its impact compared to printed media and broadening the scope of understanding media beyond a simplistic aspect of media such as one delivery platform (e.g. the World Wide Web) aside from many other options. The term is also relevant to professional career development regarding related skill set.

Primary Uses

• Journalism	• Fine art	• Entertainment
o News	o Digital art	o Movies
• Commerce	o Digital photography	o Music
o Marketing	o Experimental music	o Television
o Advertising	o Video	o Video games
o Barker channel	• Industry	• Government
o Digital signage	o Corporate communications	o Infrastructure
o Graphic design	o Business presentations	☐ Communications
• Education	o Telecommuting	☐ Transportation
o Professional training	• Software interfaces	☐ Public services
• Science	• Computer simulations	☐ Community bulletin board
• Engineering	• Virtual reality	o Military
		• Nonprofit services

Broadcasting

Broadcasting antenna in Stuttgart

Broadcasting is the distribution of audio and/or video content or other messages to a dispersed audience via any electronic mass communications medium, but typically one using the electromagnetic spectrum (radio waves), in a one-to-many model. Broadcasting began with AM radio, which came into popular use around 1920 with the spread of vacuum tube radio transmitters and receivers. Before this, all forms of electronic communication (early radio, telephone, and telegraph) were one-to-one, with the message intended for a single recipient. The term *broadcasting*, borrowed from the agricultural method of sowing seeds in a field by casting them broadly about, was coined by either KDKA manager Frank Conrad or RCA historian George Clark around 1920 to distinguish this new activity of "one-to-many" communication; a single radio station transmitting to multiple listeners.

Over the air broadcasting is usually associated with radio and television, though in recent years both radio and television transmissions have begun to be distributed by cable (cable television). The receiving parties may include the general public or a relatively small subset; the point is that anyone with the appropriate receiving technology and equipment (e.g., a radio or television set) can receive the signal. The field of broadcasting includes both government-managed services such as public radio, community radio and public television, and private commercial radio and commercial television. The U.S. Code of Federal Regulations, title 47, part 97 defines "broadcasting" as "transmissions intended for reception by the general public, either direct or relayed". Private or two-way telecommunications transmissions do not qualify under this definition. For example, amateur ("ham") and citizens band (CB) radio operators are not allowed to broadcast. As defined, "transmitting" and "broadcasting" are not the same.

Transmission of radio and television programs from a radio or television station to home receivers by radio waves is referred to as "over the air" (OTA) or terrestrial broadcasting and in most countries requires a broadcasting license. Transmissions using a wire or cable, like cable television (which also retransmits OTA stations with their consent), are also considered broadcasts, but do not necessarily require a license (though in some countries, a license is required). In the 2000s, transmissions of television and radio programs via streaming digital technology have increasingly been referred to as broadcasting as well, though strictly speaking this is incorrect.

History

The earliest broadcasting consisted of sending telegraph signals over the airwaves, using Morse code, a system developed in the 1830s by Samuel F. B. Morse, physicist Joseph Henry and Alfred Vail. They developed an electrical telegraph system which sent pulses of electric current along wires which controlled an electromagnet that was located at the receiving end of the telegraph system. A code was needed to transmit natural language using only these pulses, and the silence between them. Morse therefore developed the forerunner to modern International Morse code. This was particularly important for ship-to-ship and ship-to-shore communication, but it became increasingly important for business and general news reporting, and as an arena for personal communication by radio amateurs (Douglas, op. cit.). Audio broadcasting began experimentally in the first decade of the 20th century. By the early 1920s radio broadcasting became a household medium, at first on the AM band and later on FM. Television broadcasting started experimentally in the 1920s and became widespread after World War II, using VHF and UHF spectrum. Satellite broadcasting was initiated in the 1960s and moved into general industry usage in the 1970s, with DBS (Direct Broadcast Satellites) emerging in the 1980s.

Originally all broadcasting was composed of analog signals using analog transmission techniques but in the 2000s, broadcasters have switched to digital signals using digital transmission. In general usage, broadcasting most frequently refers to the transmission of information and entertainment programming from various sources to the general public.

- Analog audio vs. HD Radio
- Analog television vs. Digital television
- Wireless

The world's technological capacity to receive information through one-way broadcast networks more than quadrupled during the two decades from 1986 to 2007, from 432 exabytes of (optimally compressed) information, to 1.9 zettabytes. This is the information equivalent of 55 newspapers per person per day in 1986, and 175 newspapers per person per day by 2007.

Methods

Historically, there have been several methods used for broadcasting electronic media audio and/or video to the general public:

- Telephone broadcasting (1881–1932): the earliest form of electronic broadcasting (not counting data services offered by stock telegraph companies from 1867, if ticker-tapes are excluded from the definition). Telephone broadcasting began with the advent of Théâtrophone ("Theatre Phone") systems, which were telephone-based distribution systems allowing subscribers to listen to live opera and theatre performances over telephone lines, created by French inventor Clément Ader in 1881. Telephone broadcasting also grew to include telephone newspaper services for news and entertainment programming which were introduced in the 1890s, primarily located in large European cities. These telephone-based subscription services were the first examples of electrical/electronic broadcasting and offered a wide variety of programming.
- Radio broadcasting (experimentally from 1906, commercially from 1920); audio signals sent through the air as radio waves from a transmitter, picked up by an antenna and sent to a receiver. Radio stations can be linked in radio networks to broadcast common radio programs, either in broadcast syndication, simulcast or subchannels.
- Television broadcasting (telecast), experimentally from 1925, commercially from the 1930s: an extension of radio to include video signals.
- Cable radio (also called "cable FM", from 1928) and cable television (from 1932): both via coaxial cable, originally serving principally as transmission media for programming produced at either radio or television stations, but later expanding into a broad universe of cable-originated channels.
- Direct-broadcast satellite (DBS) (from c. 1974) and satellite radio (from c. 1990): meant for direct-to-home broadcast programming (as opposed to studio network uplinks and downlinks), provides a mix of traditional radio or television broadcast programming, or both, with dedicated satellite radio programming.

- Webcasting of video/television (from c. 1993) and audio/radio (from c. 1994) streams: offers a mix of traditional radio and television station broadcast programming with dedicated Internet radio and Internet television.

Economic Models

There are several means of providing financial support for continuous broadcasting:

- Commercial broadcasting: for-profit, usually privately owned stations, channels, networks, or services providing programming to the public, supported by the sale of time to advertisers for radio or television advertisements during or in breaks between programs, often in combination with cable or pay cable subscription fees.
- Public broadcasting: usually non-profit, publicly owned stations or networks supported by license fees, government funds, grants from foundations, corporate underwriting, and audience memberships and/or contributions, or a combination of these.
- Community broadcasting

Broadcasters may rely on a combination of these business models. For example, in the United States, National Public Radio (NPR) and the Public Broadcasting Service (PBS, television) supplement public membership subscriptions and grants with funding from the Corporation for Public Broadcasting (CPB), which is allocated bi-annually by Congress. US public broadcasting corporate and charitable grants are generally given in consideration of underwriting spots which differ from commercial advertisements in that they are governed by specific FCC restrictions, which prohibit the advocacy of a product or a "call to action".

Recorded and Live Forms

A television studio production control room in Olympia, Washington, August 2008.

The first regular television broadcasts started in 1937. Broadcasts can be classified as "recorded" or "live". The former allows correcting errors, and removing superfluous or undesired material, rearranging it, applying slow-motion and repetitions, and other techniques to enhance the program. However, some live events like sports television can include some of the aspects including slow-motion clips of important goals/hits, etc., in between the live television telecast. American radio-network broadcasters habitually forbade prerecorded broadcasts in the 1930s and 1940s requiring radio programs played for the Eastern and Central time zones to be repeated three hours later for the Pacific time zone. This restriction was dropped for special occasions, as in the case of the German dirigible airship *Hindenburg* disaster at Lakehurst, New Jersey, in 1937. During World War II, prerecorded broadcasts from war correspondents were allowed on U.S. radio. In addition, American radio programs were recorded for playback by Armed Forces Radio radio stations around the world.

A disadvantage of recording first is that the public may know the outcome of an event from another source, which may be a "spoiler". In addition, prerecording prevents live radio announcers from deviating from an officially approved script, as occurred with propaganda broadcasts from Germany in the 1940s and with Radio Moscow in the 1980s. Many events are advertised as being live, although they are often "recorded live" (sometimes called "live-to-tape"). This is particularly true of performances of musical artists on radio when they visit for an in-studio concert performance. Similar situations have occurred in television production ("*The Cosby Show* is recorded in front of a live television studio audience") and news broadcasting.

A broadcast may be distributed through several physical means. If coming directly from the radio studio at a single station or television station, it is simply sent through the studio/transmitter link to the transmitter and hence from the television antenna located on the radio masts and towers out to the world. Programming may also come through a communications satellite, played either live or recorded for later transmission. Networks of stations may simulcast the same programming at the same time, originally via microwave link, now usually by satellite. Distribution to stations or networks may also be through physical media, such as magnetic tape, compact disc (CD), DVD, and sometimes other formats. Usually these are included in another broadcast, such as when electronic news gathering (ENG) returns a story to the station for inclusion on a news programme.

The final leg of broadcast distribution is how the signal gets to the listener or viewer. It may come over the air as with a radio station or television station to an antenna and radio receiver, or may come through cable television or cable radio (or "wireless cable") via the station or directly from a network. The Internet may also bring either internet radio or streaming media television to the recipient, especially with multicasting allowing the signal and bandwidth to be shared. The term "broadcast network" is often used to distinguish networks that broadcast an over-the-air television signals that can be received using a tuner (television) inside a television set with a television antenna from so-called networks that are broadcast only via cable television (cablecast) or satellite television that uses a dish antenna. The term "broadcast television" can refer to the television programs of such networks.

Social Impact

The sequencing of content in a broadcast is called a schedule. As with all technological endeavors, a number of technical terms and slang have developed. A list of these terms can be found at List of broadcasting terms. Television and radio programs are distributed through radio broadcasting or cable, often both simultaneously. By coding signals and having a cable converter box with decoding equipment in homes, the latter also enables subscription-based channels, pay-tv and pay-per-view services. In his essay, John Durham Peters wrote that communication is a tool used for dissemination. Durham stated, "Dissemination is a lens—sometimes a usefully distorting one—that helps us tackle basic issues such as interaction, presence, and space and time...on the agenda of any future communication theory in general" (Durham, 211). Dissemination focuses on the message being relayed from one main source to one large audience without the exchange of dialogue in between. It is possible for the message to be changed or corrupted by government officials once the main source releases it. There is no way to predetermine how the larger population or audience will absorb the message. They can choose to listen, analyze, or simply ignore it. Dissemination in communication is widely used in the world of broadcasting.

Radio station WTUL studio, Tulane University, New Orleans

Broadcasting focuses on getting a message out and it is up to the general public to do what they wish with it. Durham also states that broadcasting is used to address an open-ended destination (Durham, 212). There are many forms of broadcasting, but they all aim to distribute a signal that will reach the target audience. Broadcasters typically arrange audiences into entire assemblies (Durham, 213). In terms of media broadcasting, a radio show can gather a large number of followers who tune in every day to specifically listen to that specific disc jockey. The disc jockey follows the script for his or her radio show and just talks into the microphone. He or she does not expect immediate feedback from any listeners. The message is broadcast across airwaves throughout the community, but there the listeners cannot always respond immediately, especially since many radio shows are recorded prior to the actual air time.

References

- Douglas, Susan J. (1987). Inventing American Broadcasting, 1899-1922. Johns Hopkins University Press. ISBN 9780801838323.
- Greb, Gordon; Adams, Mike (2003). Charles Herrold, Inventor of Radio Broadcasting. McFarland. pp. 220–221. ISBN 0786483598.
- Schlosberg, Justin (2011). "Why Does Illegal Broadcasting Continue To Thrive In The Age Of Spectrum Liberalization?". Academic Search Premier: 7. Retrieved 19 February 2013.
- "The State of the News Media 2013: An Annual Report in American Journalism," the Pew Research Center's Project for Excellence in Journalism, 2 May 2012. Page retrieved 23 May 2013.
- "The State of the News Media 2013: An Annual Report in American Journalism published by the Pew Research Center's Project for Excellence in Journalism," the Pew Research Center's Project for Excellence in Journalism, 2 May 2012. Page retrieved 23 May 2013.

- "The Elements of Journalism: What Newspeople Should Know and the Public Should Expect – Introduction Project for Excellence in Journalism (PEJ)". Journalism.org. 2006-06-19. Archived from the original on 2 October 2013. Retrieved 2013-02-23.
- "ASNE (American Society of Newspapers Editors) – Statement of Principles". Web.archive.org. Archived from the original on 5 June 2008. Retrieved 2013-03-01.
- "APME (Associated Press Managing Editors) – Statement of Ethical Principles". Web.archive.org. 2008-06-22. Archived from the original on 22 June 2008. Retrieved 2013-03-01.
- Corcoran, Mark (21 February 2012). "Drone journalism takes off". Australian Broadcasting Corporation. Retrieved 25 March 2012.

2

Broadcast Engineering: An Overview

Broadcast engineering deals exclusively with radio and television broadcasting. The chapter provides a detailed summary about broadcast engineering, its importance, functions and its scope. There is also a section about the skills and knowledge that broadcast engineers are generally required to have. Electronic media and broadcasting is best understood in confluence with the major topics listed in the following chapter.

Broadcast engineering is the field of electrical engineering, and now to some extent computer engineering and information technology, which deals with radio and television broadcasting. Audio engineering and RF engineering are also essential parts of broadcast engineering, being their own subsets of electrical engineering.

Broadcast engineering involves both the studio end and the transmitter end (the entire airchain), as well as remote broadcasts. Every station has a broadcast engineer, though one may now serve an entire station group in a city, or be a contract engineer who essentially freelances his or her services to several stations (often in small media markets) as needed.

Duties

Modern duties of a broadcast engineer include maintaining broadcast automation systems for the studio and automatic transmission systems for the transmitter plant. There are also important duties regarding radio towers, which must be maintained with proper lighting and painting. Occasionally a station's engineer must deal with complaints of RF interference, particularly after a station has made changes to its transmission facilities.

Titles

Broadcast engineers may have varying titles depending on their level of expertise and field specialty. Some widely used titles include:

- Broadcast design engineer
- Broadcast Integration Engineer
- Broadcast systems engineer
- Broadcast IT engineer
- Broadcast IT systems engineer
- Broadcast network engineer
- Broadcast maintenance engineer
- Video broadcast engineer

- TV studio broadcast engineer
- Outside broadcast engineer
- Remote broadcast engineer

Qualifications

Broadcast engineers may need to possess some or all of the following degrees, depending on the broadcast technical environment. If one of the formal qualifications is not present, a related degree or equivalent professional experience is desirable.

- Degree in electrical engineering
- Degree in electronic engineering
- Degree in telecommunications engineering
- Degree in computer engineering
- Degree in management information system
- Degree in broadcast technology

Knowledge

Broadcast engineers are generally required to have knowledge in the following areas, from conventional video broadcast systems to modern Information Technology:

- Conventional broadcast
  - Audio/Video instrumentation measurement
  - Baseband video – standard / high-definition
  - Broadcast studio acoustics
  - Television studios - broadcast video cameras and camera lenses
  - Production switcher (Video mixer)
  - Audio mixer
- Broadcast IT
  - Video compression - DV25, MPEG, DVB or ATSC (or ISDB)
  - Digital server playout technologies. - VDCP, Louth, Harris, control protocols
  - Broadcast automation
  - Disk storage – RAID / NAS / SAN technologies.
  - Archives – Tape archives or grid storage technologies.
  - Computer networking

    - Operating systems – Microsoft Windows / Mac OS / Linux / RTOS
    - Post production – video capture and non-linear editing systems (NLEs).
- RF
    - RF satellite uplinking – High-powered amplifiers (HPA)
    - RF communications satellite downlinking – Band detection, carrier detection and IRD tuning, etc.
    - RF transmitter maintenance - IOT UHF transmitters, solid state VHF transmitters, solid state MF transmitters (AM radio), tube type VHF and MF transmitters. antennas, transmission line, high power filters, digital modulators, towers, tower lighting systems, backup generators.
- Health and safety
    - Occupational safety and health
    - Fire suppression systems like FM 200.
    - Basic structural engineering
    - RF hazard mitigation

Above mentioned requirements vary from station to station.

Digital Engineering

The conversion to digital broadcasting means broadcast engineers must now be well-versed in digital television and digital radio, in addition to analogue principles. New equipment from the transmitter to the radio antenna to the receiver may be encountered by engineers new to the field. Furthermore, modern techniques place a greater demand on an engineer's expertise, such as sharing broadcast towers or radio antennas among different stations (diplexing).

Digital audio and digital video have revolutionized broadcast engineering in many respects. Broadcast studios and control rooms are now already digital in large part, using non-linear editing and digital signal processing for what used to take a great deal of time or money, if it was even possible at all. Mixing consoles for both audio and video are continuing to become more digital in the 2000s, as is the computer storage used to keep digital media libraries. Effects processing and TV graphics can now be realized much more easily and professionally as well.

Other devices used in broadcast engineering are telephone hybrids, broadcast delays, and dead air alarms.

Engineering Services

Broadcast stations often call upon outside engineering services for certain needs. For example, because structural engineering is generally not a direct part of broadcast engineering, tower companies usually design broadcast towers.

Other companies specialize in both broadcast engineering and broadcast law, which are both essential when making an application to a national broadcasting authority for a construction permit or broadcast license. This is especially critical in North America, where stations bear the entire burden of proving that their proposed facilities will not cause interference and are the best use of the radio spectrum. Such companies now have special software that can map projected radio propagation and terrain shielding, as well as lawyers that will defend the applications before the U.S. Federal Communications Commission (FCC), Canadian Radio-television and Telecommunications Commission (CRTC), or the equivalent authorities in some other countries.

Organizations

Brazil

- SET – Sociedade Brasileira de Engenharia de Televisão e Telecomunicações, Brazilian Society of Television and Telecommunications Engineering, Rio de Janeiro, Brazil.

Canada

- Central Canada Broadcast Engineers (CCBE), Paris, Ontario, Canada
- Western Association of Broadcast Engineers (WABE), Victoria, British Columbia, Canada
- IEEE Broadcast Technology Society (BTS) – Ottawa Section, Canada

Germany

- FKTG – Fernseh- und Kinotechnische Gesellschaft e.V., Germany

India

- Broadcast Engineering Society – BES (India), New Delhi, India

Iran

- Iran Broadcasting University Iran, Tehran (plus two campuses in Qom and Dubai)

Japan

- Institute of Electrical and Electronics Engineers - Japan Council

Mexico

- Asociación Mexicana de Ingenieros y Técnicos en Radiodifusión A.C. (AMITRA), Delegación Benito Juárez, México

People's Republic of China (PRC)

- China Society of Motion Picture and Television Engineers, China
- IEEE Broadcast Technology Society (BTS) – Beijing Section, China
- IEEE Broadcast Technology Society (BTS) – Shanghai Section, China

People's Republic of China – Hong Kong

- Society of Motion Picture and Television Engineers-(Hong Kong) Section, 電影電視工程師協會香港分會 Hong Kong
- Society of Broadcast Engineers Hong Kong Chapter, 廣播工程師協會香港分會 Hong Kong
- Hong Kong Televisioners Association (HKTVA), 香港電視專業人員協會 Hong Kong

Philippines

- Society of Broadcast Engineers and Technicians of the Philippines, Inc. (SBETP), Quezon City, Philippines

Republic of China – Taiwan

- IEEE Broadcast Technology Society (BTS) – Taipei Section

South Korea

- Korean Broadcast Engineers & Technicians Association (KOBETA), Seoul, South Korea (ROK)

Turkey

- Chamber of Electrical Engineers (EMO) (Joint chamber of electrical, electronics and bio-medical engineers)

United States

In the United States, many broadcast engineers belong to the Society of Broadcast Engineers (SBE). Some may also belong to the Society of Motion Picture and Television Engineers (SMPTE), or to organizations of related fields, such as the Audio Engineering Society or Institute of Electrical and Electronics Engineers (IEEE) - IEEE Broadcast Technology Society (BTS).

For public radio, the Association of Public Radio Engineers was created in late May 2006.

Uruguay

- ANDEBU – Asociación Nacional de Broadcasters Uruguayos, Montevideo, Uruguay

References

- "about.com - Broadcast Technician or Sound Engineering Technician: Career Information". Careerplanning.about.com. 2010-11-08. Retrieved 2013-08-03.
- "U.S. Department of Labor - Bureau of Labor Statistics - Broadcast and Sound Engineering Technicians and Radio Operators". Bls.gov. 2012-03-29. Retrieved 2013-08-03.
- "Society of Motion Picture and Television Engineers-(Hong Kong) Section (SMPTE-HKG)". Smpte.org.hk. Retrieved 2013-08-03.

3

Evolution of Broadcasting

Broadcasting refers to the transmission of audio or video to an audience over electronic mass-media devices like radio, television and computers. This chapter chronicles the history of mass-media broadcasting and informs the reader about the progress that has been till date in the field. The reader will gain a deep understanding of this field.

History of Broadcasting

The first broadcasting of a radio transmission consisted of Morse code (or wireless telegraphy) was made from a temporary station set up by Guglielmo Marconi in 1895. This followed on from pioneering work in the field by Alessandro Volta, André-Marie Ampère, Georg Ohm, James Clerk Maxwell and Heinrich Hertz. The broadcasting of music and talk via radio started experimentally around 1905-1906, and commercially around 1920 to 1923. VHF (very high frequency) stations started 30 to 35 years later.

Guglielmo Marconi.

In the early days, radio stations broadcast on the long wave, medium wave and short wave bands, and later on VHF (very high frequency) and UHF (ultra high frequency). However, in the United Kingdom, Hungary, France and some other places, from as early as 1890 there was already a system whereby news, music, live theatre, music hall, fiction readings, religious broadcasts, etc., were available in private homes [and other places] via the conventional telephone line, with subscribers being supplied with a number of special, personalised headsets. In Britain this system was known as

Electrophone, and was available as early as 1895 or 1899 [sources vary] and up until 1926. In Hungary, it was called Telefon Hírmondó [1893-1920s], and in France, Théâtrophone [1890-1932]). The Wikipedia Telefon Hírmondó page includes a 1907 program guide which looks remarkably similar to the types of schedules used by many broadcasting stations some 20 or 30 years later.

The Marconi Company was formed in England in 1897. The photo shows a typical early scene, from 1906, with Marconi empoyee Donald Manson at right.

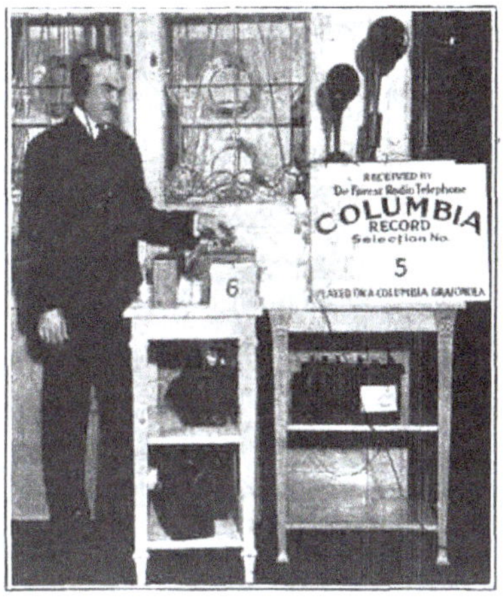

Lee DeForest broadcasting Columbia phonograph records on pioneering New York station 2XG,in 1916.

The British Broadcasting Corporation's landmark and iconic London headquarters, Broadcasting House, opened in 1932. At right is the 2005 eastern extension, the John Peel wing.

By the 1950s, virtually every country had a broadcasting system, typically one owned and operated by the government. Alternative modes included commercial radio, as in the United States; or a dual system with both state sponsored and commercial stations, introduced in Australia as early as 1924, with Canada following in 1932. Today, most countries have evolved into a dual system, including the UK. By 1955, practically every family in North America and Western Europe, as well as Japan, had a radio. A dramatic change came in the 1960s with the introduction of small inexpensive portable transistor radio, the greatly expanded ownership and usage. Access became practically universal across the world. Broadcasting has seen many improvements, refinements and challenges; these include (but are not confined to):

- international broadcasts, particularly on short wave band;
- better technology, which saw radios becoming cheaper, and available in almost every home, as well as in cars and portable sets;
- the introduction of FM broadcasting and its effect on AM stations;
- the challenge of television, which meant that radio broadcasters later concentrated on music of varying types, news, sport and discussion programs;
- the invention of the transistor, meaning even greater portability and even cheaper radio and TV sets;
- digital radio;
- Internet
- internet radio.

Early Broadcasting

Australia

The History of broadcasting in Australia has been shaped for over a century by the problem of communication across long distances, coupled with a strong base in a wealthy society with a deep taste for aural communications. Australia developed its own system, through its own engineers, manufacturers, retailers, newspapers, entertainment services, and news agencies. The government set up the first radio system, and business interests marginalized the hobbyists and amateurs. The Labor Party was especially interested in radio because it allowed them to bypass the newspapers, which were mostly controlled by the opposition. Both parties agreed on the need for a national system, and in 1932 set up the Australian Broadcasting Commission, as a government agency that was largely separate from political interference. The first commercial broadcasters, originally known as "B" class stations, were on the air as early as 1925. The number of stations (commercial and national) remained relatively dormant throughout World War II and in the post-war era. However, in the 1970s, the Labor government under Prime Minister Gough Whitlam commenced a broadcasting renaissance so that by the 1990s there were 50 different radio services available for groups based on tastes, languages, religion, or geography. The broadcasting system was largely deregulated in 1992, except that there were limits on foreign ownership and on monopolistic control. By 2000, 99 percent of Australians owned at least one television set, and averaged 20 hours a week watching it.

Formative Years

Australian radio hams can be traced to the early 1900s. The 1905 *Wireless Telegraphy Act* whilst acknowledging the existence of wireless telgraphy, brought all broadcasting matters in Australia under the control of the Federal Government. In 1906, the first official Morse code transmission in Australia was by the Marconi Company between Queenscliff, Victoria and Devonport, Tasmania.

Experiments with Broadcasting Music

The first broadcast of music was made during a demonstration on 13 August 1919 by Ernest Fisk (later Sir Ernest) of AWA – Amalgamated Wireless (Australasia). A number of amateurs commenced broadcasting music in 1920 and 1921. These included 2CM, Sydney; 2YG, Sydney; 2XY, Newcastle; 3ME, Melbourne; 3DP, Melbourne; 4CM, Brisbane; 4AE, Brisbane; 4CH, Brisbane; 5AC, Adelaide; 5AD, Adelaide (not associated with 5AD which commenced in 1930); 5BG, Adelaide; 7AA, Hobart; 7AB, Hobart. Many other amateurs soon followed. 2CM was run by Charles MacLuran who started the station in 1921 with regular Sunday evening broadcasts from the Wentworth Hotel, Sydney. 2CM is often regarded as Australia's first, regular, non-official station.

Sealed Set System

It was not until November 1923 when the government finally gave its approval for a number of officially recognised medium wave stations. These were (with the dates they came on air):

- 2SB, Sydney, Sydney Broadcasters Ltd, 23 November 1923 (known as 2BL from 1 March 1924);
- 2FC, Sydney, Farmers & Co Ltd, 8 December 1923;
- 3AR, Melbourne,Associated Radio Co, 26 January 1924;
- 3LO, Melbourne, Broadcasting Co of Australia, 13 October 1924;
- 6WF, Perth, Westralian Farmers, 4 June 1924.

All stations operated under a unique Sealed Set system under which each set was sealed to the frequency of one station. Part of the price of the set went to the government via the Postmaster-General's Department (PMG), with money also going to the broadcaster. Apart from extremely limited advertising, this was the broadcasters' only source of income. From the outset problems with the system came to the fore. Many young people built their own sets, which could receive all the stations. The sealed set system was devised by broadcasting pioneer Ernest Fisk of AWA – Amalgamated Wireless (Australasia).

Categories in Australia from 1924

As quickly as July 1924, the Sealed Set system was declared to be unsuccessful and it was replaced by a system of A Class and B Class stations. There were one or two A Class stations in each major market and these were paid for by a listener's licence fee imposed on all listeners-in. The five former sealed set stations became A Class stations, and they were soon joined by the following stations in other State capitals:

- 5CL, Adelaide, Central Broadcasters Ltd, 20 November 1924;
- 7ZL, Hobart, Associated Radio Co, 17 December 1924;
- 4QG, Brisbane, Queensland Radio Service (operated by the Queensland government), 27 July 1925.

Emil Voigt, founder of 2KY on behalf of the Labor Council of New South Wales. This photo was taken in earlier days when Voight was a prominent British athlete, and winner of the Gold Medal for the five mile race at the 1908 Summer Olympics in London.

From 1929, all A Class stations received all their programs from the one source, the Australian Broadcasting Company which was made up of the following shareholders: Greater Union Theatres (a movie theatre chain), Fuller's Theatres (a live theatre chain) and J. Albert & Sons (music publishers and retailers). A number of B Class stations were also licensed. These did not receive any government monies and were expected to derive their income from advertising, sponsorship, or other sources. Within a few years B Class stations were being referred to as "commercial stations". The following were the first to be licensed:

- 2BE, Sydney, 7 November 1924 (closed 6 November 1929);
- 2EU, Sydney, 26 January 1925, still on the air – name changed to 2UE within months of opening;
- 2HD, Newcastle, 27 January 1925, still on the air;
- 2UW, Sydney, 13 February 1925, still on the air (date needs to be confirmed);
- 5DN, Adelaide, 24 February 1925, still on the air;
- 3WR, Wangaratta, 25 February 1925 (closed 25 November 1925 but later re-opened);
- 3UZ, Melbourne, 8 March 1925, still on the air;
- 4GR, Toowoomba, 16 August 1925, still on the air;

- 2GB, Sydney, 23 August 1926, still on the air.
- 2KY, Sydney, 31 October 1925, still on the air;
- 2MK, Bathurst, 31 October 1925 (closed November 1931);

ABC mobile studio caravan, used for concerts presented by the ABC at army camps and other locations, 1940

Amateur broadcasters continued to operate in the long wave and short wave bands. In Melbourne, for some years, they were also permitted to broadcast on the medium wave band on Sundays between 12.30 and 2.30 pm, during which time all commercial stations were required to close down. In the late 1920s and early 1930s, the PMG planned to institute C Class stations which would have had their advertising limited to the station owner(s) only. When the plan was abandoned in 1931, the PMG was about to issue such a licence to the Akron Tyre Co in Melbourne; in lieu of a C Class licence, Akron was given a licence for a B Class station but with a number of limiting conditions on its licence. A national service, the Australian Broadcasting Commission, was formed in July 1932, when the Australian Broadcasting Company's contract expired. The Corporation took over the assets of all A Class stations. It still exists as the Australian Broadcasting Corporation. The Australian Broadcasting Co changed its name to the Commonwealth Broadcasting Company and later the Australian Radio Network. It soon purchased Sydney commercial station 2UW and now has an Australia-wide network of commercial stations.

Types of Programs

As with most countries, most Australian stations originally broadcast music interspersed with such things as talks, coverage of sporting events, church broadcasts, weather, news and time signals of various types. Virtually all stations also had programs of interest to women, and children's sessions. From the outset, A Class stations' peak-hour evening programs often consisted of live broadcasts from various theatres, i.e. dramas, operas, musicals, variety shows, vaudeville, etc. The first dramas especially written for radio were transmitted in the mid-1920s. By the 1930s, the ABC was transmitting a number of British programs sourced from the BBC, and commercial stations were receiving a number of US programs, particularly dramas. However, in the 1940s, war-time restrictions made it difficult to access overseas programs and, therefore, the amount of Australian dramatic material increased. As well as using original ideas and scripts, there were a number of local versions of overseas programs. Initially, much of the music broadcast in Australia was from live studio concerts. However, the amount of gramophone (and piano roll) music soon increased dramatically, particularly on commercial stations.

In the late 1930s, the number of big production variety shows multiplied significantly, particularly on the two major commercial networks, Macquarie and Major. After World War II the independent Colgate-Palmolive radio production unit was formed. It poached most major radio stars from the various stations. Until the 1950s, the popular image of the whole family seated around a set in the living room was the most accepted way of listening to radio. Therefore, most stations had to be all things to all people, and specialised programming was not really thought about at this stage (it did not come in until the late 1950s). Because of this, programming on most stations was pretty much the same.

In the immediate post-war period, nearly all commercial stations had a schedule that looked something like this: a breakfast session with bright music including band music, news and weather, a children's segment and, usually, an exercise segment; morning programs were aimed at women listeners, often with large blocks of soap operas or serials, and many segments of the *handy hints* genre; afternoon programs were also usually geared at women but with more music and, often, a request session; after school, there was inevitably a Children's Session (often hosted by an *aunt* or *uncle* or both), and usually featuring birthday greetings; this was followed by another block of serials, often geared at children, and/or dinner music; the major news bulletin was usually at 7.00 pm, often followed by a news commentary; the peak listening hours typically consisted of a mix of variety programs (including many quizzes), dramas, talent quests and the occasional musical program, often live; late night programing mainly consisted of relaxing music, usually mellow jazz or light classical. There was usually only one station in each capital city that was licensed to broadcast through the night.

Early Experiments with Television

As early as 1929, two Melbourne commercial radio stations, 3UZ and 3DB were conducting experimental mechanical television broadcasts – these were conducted in the early hours of the morning, after the stations had officially closed down. In 1934 Dr Val McDowall at amateur station 4CM Brisbane conducted experiments in electronic television.

Mobile Stations

Two of Australia's most unusual medium wave stations were mobile stations 2XT and 3YB. They both operated in eras prior to the universal establishment of rural radio stations. 2XT was designed and operated by AWA within the State of New South Wales, from a NSW Railways train, between November 1925 and December 1927. 2XT, which stood for *experimental train*, visited over 100 rural centres. Engineers would set up a transmitting aerial and the station would then begin broadcasting. This led to the further sales of AWA products. 3YB provided a similar service in rural Victoria between October 1931 and November 1935. Initially, the station operated from a Ford car and a Ford truck, but from 17 October 1932 they operated from a converted 1899 former Royal Train carriage. Whilst the engineers were setting up the station's 50-watt transmitter in the town being visited, salesmen would sign up advertisers for the fortnight that 3YB would broadcast from that region. The station was on the air from 6.00 and 10.00 pm daily, and its 1,000-record library was divided into set four-hour programs, one for each of 14 days. In other words, the music broadcast from each town was identical. The station was operated by Vic Dinenny, but named after announcer Jack Young from Ballarat. On 18 January 1936, Dinenny set up 3YB Warnambool, followed on 18 May 1937 by 3UL Warragul.

The merchant vessel MV Kanimbla is believed to be the world's only ship built with an inbuilt broadcasting station. The Kanimlba was constructed in Northern Ireland in 1936 and was primarily designed for McIlwraith McEachern Limited to ply passengers between Cairns, Queensland and Fremantle, Western Australia. The broadcasting station was constructed and operated by AWA and was initially given the ham radio callsign VK9MI but was later 9MI. (At this time, the "9" in the callsign was aberrationary). The station made an experimental broadcast before leaving Northern Ireland, and a number of such broadcasts at sea, on the way to Australia. 9MI's first official broadcast in April 1939 was made from the Great Australian Bight. The station broadcast on short wave, usually a couple of times per week, but many of its programs were relayed to commercial medium wave stations that were also owned by AWA. The 9MI manager and announcer (and probably the only member of staff) was Eileen Foley. 9MI ceased broadcasting at the commencement of World War II in September 1939. The Kanimbla was commissioned as a Royal Navy (later Royal Australian Navy) vessel with the name HMS/HMAS Kanimbla. It had an extremely prominent and successful war-time career.

Canada

The history of broadcasting in Canada begins as early as 1919 with the first experimental broadcast programs in Montreal. Canadians were swept up in the radio craze and built crystal sets to listen to American stations while The Marconi Wireless Telegraph Company of Canada offered its first commercially produced radio-broadcast receiver (Model "C") in 1921, followed by its "Marconiphone" Model I in 1923. Main themes in the history include the development of the engineering technology; the construction of stations across the country and the building of networks; the widespread purchase and use of radio and television sets by the general public; debates regarding state versus private ownership of stations; financing of the broadcasts media through the government, license fees, and advertising; the changing content of the programming; the impact of the programming on Canadian identity; the media's influence on shaping audience responses to music, sports and politics; the role of the Québec government; Francophone versus Anglophone cultural tastes; the role of other ethnic groups and First Nations; fears of American cultural imperialism via the airwaves; and the impact of the Internet and smartphones on traditional broadcasting media.

Radio signals carried long distances, and a number of American stations could easily be received in parts of Canada. The first Canadian station was CFCF, originally an experimental station from the Marconi Company in Montreal. Civilian use of Wireless Telegraphy had been forbidden in Canada for the duration of World War I. The Marconi Wireless Telegraph Company of Canada was the only one to retain the right to continue radio experiments for military use. This proved instrumental in giving the company a lead in developing an experimental radio broadcasting station immediately after the war. The first radio broadcast in Canada was accomplished by The Marconi Wireless Telegraph Company of Canada in Montreal on December 1, 1919 under the call sign XWA (for

"Experimental Wireless Apparatus") from its Williams Street factory. The station began regular programming on May 20, 1920 and its call letters were changed to CFCF on November 4, 1920. In Toronto, the first radio station was operated by the *Toronto Star* newspaper. Station CKCE began in April 1922. and were so well received that the Star pushed forward with its own studios and transmitting facilities, returning to the air as CFCA in late June 1922. In Montreal, another newspaper, *La Presse*, put its own station, CKAC on the air in late September 1922. Because there were governmental limitations on radio frequencies back then, CKAC and CFCF alternated—one would broadcast one night, and the other would broadcast the night after that. For a time, CKAC was broadcasting some programs in French, and some in English: in 1924, for example, the station rebroadcast fifteen Boston Bruins hockey games from station WBZ in Boston. Meanwhile, in other Canadian provinces, 1922 was also the year for their first stations, including CJCE in Vancouver, and CQCA (which soon became CHCQ) in Calgary.

As radio grew in popularity during the mid-1920s, a problem arose: the U.S. stations dominated the airwaves and with a limited number of frequencies available for broadcasters to use, it was the American stations that seemed to get most of them. This was despite an agreement with the US Department of Commerce (which supervised broadcasting in the years prior to the Federal Radio Commission) that a certain number of frequencies were reserved exclusively for Canadian signals. But if a US station wanted one of those frequencies, the Department of Commerce seemed unwilling to stop it, much to the frustration of Canadian owners who wanted to put stations on the air. The Canadian government and the US government began negotiations in late 1926, in hopes of finding a satisfactory solution. Meanwhile, in 1928, Canada got its first network, operated by the Canadian National Railways. CNR had already made itself known in radio since 1923, thanks in large part to the leadership of CNR's president, Sir Henry Thornton. The company began equipping its trains with radio receivers, and allowed passengers to hear radio stations from Canada and the US. In 1924, CN began building its own stations, and by 1928, it was able to create a network. In 1932, the Canadian Radio Broadcasting Commission was formed, and in 1936, the Canadian Broadcasting Corporation, the country's national radio service, made its debut.

Cuba

There was interest in radio almost from broadcasting's earliest days. Due to the proximity of Cuba to the U.S. state of Florida, some Cubans would try to listen to the American stations whose signals reached the island. But there was no radio station in Cuba until 1922. The arrival of the first radio station, PWX, was greeted with enthusiasm. PWX, owned by the Cuban Telephone Company, was located in Havana. It was a joint venture with the International Telephone and Telegraph Company of New York. PWX debuted on the air on October 10, 1922. PWX broadcast programs in both English and Spanish, and its signal was easily received at night in a number of American cities. Another early station in Cuba was owned by Frank Jones, an American amateur radio operator and Chief Engineer of the Tuinucu Sugar Company. The station used amateur call letters, and went on the air as 6KW. In late 1928, PWX began using the call letters CMC. Its slogan was "If you hear 'La Paloma,' you are in tune with CMC." As with many other countries, interest in radio expanded, and by 1932, Cuba had more than thirty stations, spread out in cities all over the island.

France

Radio Paris began operations in 1922, followed by Radio Toulouse and Radio Lyon. Before 1940,

14 commercial and 12 public sector radio stations were in operation. The government exerted tight control over radio broadcasting. Political debate was not encouraged. In the 1932 election campaign, for example, the opposition was allowed one broadcast while the incumbent made numerous campaign broadcasts. Radio was a potentially powerful new medium, but France was quite laggard in consumer ownership of radio sets, With 5 million radio receivers in 1937, compared to over 8 million and both Britain and Germany, and 26 million in the United States. The government imposed very strict controls on news dissemination. After 1938, stations were allowed only three brief daily bulletins, of seven minutes each, to cover all the day's news. The Prime Minister's office closely supervised the news items that were to be broadcast. As war approached, Frenchmen learned little or nothing about it from the radio. The government thought that policy wise, because it wanted no interference in its policies. The unexpected result, however, was the Frenchman were puzzled and uncertain great crises erupted in 1938-39, and their morale and support for government policies was much weaker than in Britain.

Germany

The first radio station in Germany went on the air in Berlin in late 1923, using the call letters "LP." Before 1933, German radio broadcasting was Conducted by 10 regional broadcasting monopolies, each of which had a government representative on its board. The Post Office Provided overall supervision. A listening fee of 2 Reichsmark per receiver paid most costs, and radio station frequencies were limited, which even restricted the number of amateur radio operators. Immediately following Hitler's assumption of power in 1933, Joseph Goebbels became head of the Ministry for Propaganda and Public Enlightenment and took full control of broadcasting. Non-Nazis were removed from broadcasting and editorial positions. Jews were fired from all positions.

Germany was easily served by a number of European mediumwave stations, including the BBC, but the Nazis made it illegal for Germans to listen to foreign broadcasts. During the war, German stations broadcast not only war propaganda and entertainment for German forces dispersed through Europe, as well as air raid alerts. There was heavy use of short wave for "Germany Calling" programmes directed at Britain and Allied forces around the world. Goebbels Also set up numerous Nazi stations that pretended to be from the Allied world. Germany experimented with television broadcasting, using a 180-line raster system beginning before 1935. German propaganda claimed the system was superior to the British mechanical scanning system, but it never became operational.

Japan

The first radio station in Japan was JOAK, which opened in Tokyo in March 1925. It was founded by Masajiro Kotamura, an inventor and engineer. It was unique in that at least one of its announcers was a woman, Akiko Midorikawa. JOAK was followed soon after by JOBK in Osaka and JOCK in Nagoya. The National Broadcasting Service, today known as NHK (Nippon Hoso Kyokai), began in August 1926. All stations were supported by licensing fees: in 1926, for example, people wishing to receive a permit to own a radio set paid a fee of one yen a month to the government. Programming on Japanese stations of the 1920s included music, news, language instruction (lessons were offered in English, French and German) and educations talks. These early stations broadcast on average about eight hours of programs a day.

Mexico

Amateur radio was very popular in Mexico; while most of the hams were male, notably Constantino de Tarnava, acknowledged in some sources as Mexico's first amateur radio operator, one of the early ham radio operators was female—Maria Dolores Estrada. But commercial radio was difficult to achieve, due to a federal regulation forbidding any broadcasts that were not for the benefit of the Mexican government. Still, in November 1923, CYL in Mexico City went on the air, featuring music (both folk songs and popular dance concerts), religious services, and news. CYL used as its slogans "El Universal" and "La Casa del Radio", and it won over the government, by giving political candidates the opportunity to use the station to campaign. Its signal was so powerful that it was even received in Canada sometimes. Pressure from listeners and potential station owners also contributed to the government relenting and allowing more stations to go on the air. In 1931, the "C" call letters were all changed to "X" call letters (XE being reserved for broadcasting), and by 1932, Mexico had nearly forty radio stations, ten of which were in Mexico City.

Philippines

Interest in amateur radio was noted in the Philippines in the early 1920s. There were radio stations operating in the Philippines, including one owned by American businessman named Henry Hermann, as early as 1922, according to some sources; not much documentation about that period of time exists. In the autumn of 1927, KZRM in Manila, owned by the Radio Corporation of the Philippines, went on the air. The Radio Corporation of the Philippines was a subsidiary of American company RCA (Radio Corporation of America). By 1932, the island had three radio stations: KRZC in Cebu, as well as KZIB (owned by a department store) and KZFM, the government-owned station in Manila. Of the stations listed by Pierre Key, KZFM was the strongest, with 50,000 watts. Two radio networks were ultimately created: one, the Manila Broadcasting Company, began as a single station, KZRH in Manila, in July 1939, and after World War II, in 1946, the station's owners began to develop their network by buying other radio properties. As for the Philippine Broadcasting Company, it too began with one station (KZFM), and received its new name in mid-1946, after the Philippines became an independent country. At the end of 1946, the new network had six stations. Both KZRH and KZFM also affiliated with American networks; the stations wanted to have access to certain popular American programs, and the American networks wanted to sell products in the Philippines.

Sri Lanka

Sri Lanka has the oldest radio station in Asia (world's second oldest). The station was known as Radio Ceylon. It developed into one of the finest broadcasting institutions in the world. It is now known as the Sri Lanka Broadcasting Corporation. Sri Lanka created broadcasting history in Asia when broadcasting was started in Ceylon by the Telegraph Department in 1923 on an experimental footing, just three years after the inauguration of broadcasting in Europe. Gramophone music was broadcast from a tiny room in the Central Telegraph Office with the aid of a small transmitter built by the Telegraph Department engineers from the radio equipment of a captured German submarine. This broadcasting experiment was successful; barely three years later, on December 16, 1925, a regular broadcasting service came to be instituted. Edward Harper who came to Ceylon as Chief Engineer of the Telegraph Office in 1921, was the first person to actively promote broadcasting in

Ceylon. Sri Lanka occupies an important place in the history of broadcasting with broadcasting services inaugurated just three years after the launch of the BBC in the United Kingdom. Edward Harper launched the first experimental broadcast as well as founding the Ceylon Wireless Club, together with British and Ceylonese radio enthusiasts on the island. Edward Harper has been dubbed ' the Father of Broadcasting in Ceylon,' because of his pioneering efforts, his skill and his determination to succeed. Edward Harper and his fellow Ceylonese radio enthusiasts, made it happen.

United Kingdom

Caricature of Sir John Reith, by the artist, Wooding.

The first experimental music broadcasts, from Marconi's factory in Chelmsford, began in 1920. Two years later, in October 1922, a consortium of radio manufacturers formed the British Broadcasting Company (BBC); they allowed some sponsored programs, although they were not what we would today consider a fully commercial station. Meanwhile, the first radio stations in England were experimental station 2MT, located near Chelmsford, and station 2LO in London: both were operated by the Marconi Company. By late 1923, there were six stations broadcasting regularly in the United Kingdom: London's 2LO, Manchester's 2ZY, and stations in Birmingham, Cardiff, Newcastle, and Glasgow. As for the consortium of radio manufacturers, it dissolved in 1926, when its license expired; it then became the British Broadcasting Corporation, a non-commercial organization. Its governors are appointed by the British government, but they do not answer to it. Lord Reith took a formative role in developing the BBC, especially in radio. Working as its first manager and Director-General, he promoted the philosophy of *public service broadcasting*, firmly grounded in the moral benefits of education and of uplifting entertainment, eschewing commercial influence and maintaining a maximum of independence from political control.

Commercial stations such as Radio Normandie and Radio Luxembourg broadcast into the UK from other European countries. This provided a very popular alternative to the rather austere BBC. These stations were closed during the War, and only Radio Luxembourg returned afterward. BBC television broadcasts in Britain began on November 2, 1936, and continued until wartime conditions closed the service in 1939.

United States

Reginald Fessenden, the *father* of radio broadcasting in the US.

Reginald Fessenden did ground-breaking experiments with voice and music by 1906. Charles "Doc" Herrold of San Jose, California sent out broadcasts as early as April 1909 from his Herrold School electronics institute in downtown San Jose, using the identification *San Jose Calling*, and then a variety of different call signs as the Department of Commerce began to regulate radio. He was on the air daily for nearly a decade when the World War interrupted operations.

"Doc" Herrold is shown at the microphone of KQW, early 1920s.

Charles Logwood broadcasting at station 2XG, New York City, *circa* November, 1916.

Pioneer radio station 2XG, also known as the "Highbridge station", was an experimental station located in New York City and licensed to the DeForest Radio Telephone and Telegraph Company. It was the first station to use a vacuum tube transmitter to make radio broadcasts on a regular schedule. From 1912 to 1917 Charles Herrold made regular broadcasts, but used an arc transmitter. He switched to a vacuum tube transmitter when he restarted broadcasting activities in 1921. Herrold coined the terms broadcasting and narrowcasting,. Herrold claimed the invention of broadcasting to a wide audience, through the use of antennas designed to radiate signals in all directions. David Sarnoff has been considered by many as "the prescient prophet of broadcasting who predicted the medium's rise in 1915", referring to his radio music box concept.

Broadcasting pioneer Frank Conrad in a 1921 portrait.

A few organizations were allowed to keep working on radio during the war. Westinghouse was the most well-known of these. Frank Conrad, a Westinghouse engineer, had been making transmissions from 8XK since 1916 that included music programming. A team at the University of Wisconsin–Madison headed by Professor Earle M. Terry was also on the air. They operated 9XM, originally licensed by Professor Edward Bennett in 1914, and experimented with voice broadcasts starting in 1917.

1920s

By 1919, after the war, radio pioneers across the country resumed transmissions. The early stations gained new call signs. Many early stations were started by newspapers worried radio might replace their newspapers. 8XK became KDKA in 1920. KDKA received the first federal license and began broadcasting on November 2, 1920. Madison Avenue early on recognized the importance of radio as a new advertising medium. Advertising provided the major funding for most stations. United States never had a licensing fee for set users. The National Broadcasting Company began regular broadcasting in 1926, with telephone links between New York and other Eastern cities. NBC became the dominant radio network, splitting into Red and Blue networks. The Columbia Broadcasting System began in 1927 under the guidance of William S. Paley.

Radio in education began as early as April 1922, when Medford Hillside's WGI Radio broadcast the first of an ongoing series of educational lectures from Tufts College professors. These lectures were described by the press as a sort of "wireless college." Soon, other colleges across the U.S. began adding radio broadcasting courses to their curricula; some, like the University of Iowa, even provided what today would be known as distance-learning credits. Curry College, first in Boston and then in Milton, Massachusetts, introduced one of the nation's first broadcasting majors in 1932 when the college teamed up with WLOE in Boston to have students broadcast programs. This success led to numerous radio courses in the curriculum which has taught thousands of radio broadcasters from the 1930s to today.

In 1934, several independent stations formed the Mutual Broadcasting System to exchange syndicated programming, including *The Lone Ranger* and *Amos 'n' Andy*. Prior to 1927, U.S. radio was supervised by the Department of Commerce. Then, the Radio Act of 1927 created the Federal Radio Commission (FRC); in 1934, this agency became known as the Federal Communications Commission (FCC). A Federal Communications Commission decision in 1939 required NBC to divest itself of its Blue Network. That decision was sustained by the Supreme Court in a 1943 decision, National Broadcasting Co. v. United States, which established the framework that the "scarcity" of radio-frequency meant that broadcasting was subject to greater regulation than other media. This Blue Network network became the American Broadcasting Company (ABC). Around 1946, ABC, NBC, and CBS began regular television broadcasts. Another TV network, the DuMont Television Network, was founded earlier, but was disbanded in 1956.

1950s and 1960s

Australia

Norman Banks was one of Melbourne's (and Australia's) most prominent broadcasters at 3KZ (1930-1952) and 3AW (1952-1978). He is remembered for founding Carols by Candlelight, as a pioneer football commentator, and for hosting both musical and interview programs. In later years he was one of Melbourne's first and most prominent talk back hosts. At the commencement of his career, Banks was known for his double entredes and risque remarks; as a talk back host he was outspoken in his conservative views, especially regarding the White Australia policy and Apartheid. In 1978 his 47-year career in radio was hailed as the longest in world history. Not including the early television experiments, mainstream television transmission commenced in Sydney and Melbourne in the latter part of 1956, that is, in time for the 1956 Melbourne Olympic Games in November/December 1956. It was then phased in in other capital cities, and then into rural markets. Many forms entertainment, particularly drama and variety, were considered more suited to television than radio, and many such programs were gradually deleted from radio schedules.

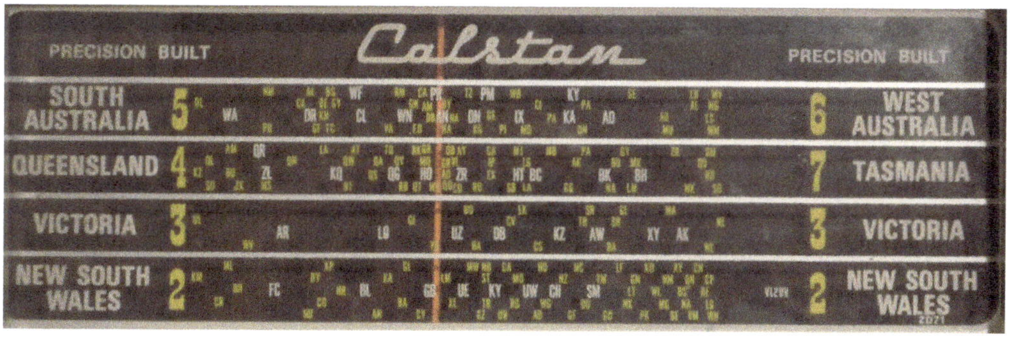

Australian radio sets usually had the positions of radio stations marked on their dials.
The illustration is a dial from a transistorised, mains operated Calstan radio, circa 1960s.
(Click image for a high resolution view, with readable callsigns.)

Naomi ("Joan") Melwit and Norman Banks at the 3KZ microphone, in late 1930s.

The transistor radio first appeared on the market in 1954. In particular, it made portable radios even more transportable. All sets quicklly became smaller, cheaper and more convenient. The aim of radio manufacturers became a radio in every room, in the car, and in the pocket. The upshot of these two changes was that stations started to specialise and concentrate on specific markets. The first areas to see specialised stations were the news and current affairs market, and stations specialising in pop music and geared toward the younger listener who was now able to afford his/her own radio. Talk back ("talk radio") became a major radio genre by the end of the 1960s, but it was not legalised in Australia until October 1967. The fears of intrusion were addressed by a beep that occurred every few seconds, so that the caller knew that his/her call was being broadcast. There was also a seven-second delay so that obscene or libelous material could be monitored. By the end of the 1960s, specialisation by radio stations had increased dramatically and there were stations focusing on various kinds of music, talk back, news, sport, etc.

Germany

When the Federal Republic of Germany was organized in 1949, its Enabling Act established strong state government powers. Broadcasting was organized on a state, rather than a national, basis. Nine regional radio networks were established. A technical coordinating organization, the Arbeitsgemeinschaft der offentlich-rechtlichen Rundfunkanstalten der Bundesrepublik Deutschland (ARD), came into being in 1950 to lessen technical conflicts. The Allied forces in Europe developed their own radio networks, including the U.S. American Forces Network (AFN). Inside Berlin, Radio in the American Sector (RIAS) became a key source of news in the German Democratic Republic. Germany began developing a network of VHF FM broadcast stations in 1955 because of the excessive crowding of the mediumwave and shortwave broadcast bands.

Sri Lanka

Radio Ceylon ruled the airwaves in the 1950s and 1960s in the Indian sub-continent. The station developed into the most popular radio network in South Asia. Millions of listeners in India for example tuned into Radio Ceylon. Announcers like Livy Wijemanne, Vernon Corea, Pearl Ondaatje, Tim Horshington, Greg Roskowski, Jimmy Bharucha, Mil Sansoni, Eardley Peiris, Shirley Perera, Bob Harvie, Christopher Greet, Prosper Fernando, Ameen Sayani (of Binaca Geetmala fame),Karunaratne Abeysekera, S.P.Mylvaganam (the first Tamil Announcer on the Commercial Service) were hugely popular across South Asia. The Hindi Service also helped build Radio Ceylon's reputation as the market leader in the Indian sub-continent. Gopal Sharma, Sunil Dutt Ameen Sayani, Hamid Sayani, were among the Indian announcers of the station. The Commercial Service of Radio Ceylon was hugely successful under the leadership of Clifford Dodd, the Australian administrator and broadcasting expert who was sent to Ceylon under the Colombo Plan. Dodd hand picked some of the most talented radio presenters in South Asia. They went on to enjoy star status in the Indian sub-continent. This was Radio Ceylon's golden era.

United Kingdom

Radio Luxembourg remained popular during the 1950s but saw its audience decline as commercial television and pirate radio, combined with a switch to a less clear frequency, began to erode its influence. BBC television resumed on June 7, 1946, and commercial television began on September 22, 1955. Both used the pre-war 405-line standard. BBC2 came on the air on April 20, 1964, using the 625-line standard, and began PAL colour transmissions on July 1, 1967, the first in Europe. The two older networks transmitted in 625-line colour from 1969. During the 1960s there was still no UK-based commercial radio. A number of 'pirate' radio ships, located in international waters just outside the jurisdiction of English law, came on the air between 1964 and 1967. The most famous of these was Radio Caroline, which was the only station to continue broadcasting after the offshore pirates were effectively outlawed on August 14, 1967 by the Marine Broadcasting Offences Act. It was finally forced off air due to a dispute over tendering payments, but returned in 1972 and continued on and off until 1990. The station still broadcasts, nowadays using satellite carriers and internet.

United States

Television began to replace radio as the chief source of revenue for broadcasting networks. Although many radio programs continued through this decade, including *Gunsmoke* and *The Guiding Light*, by 1960 networks had ceased producing entertainment programs. As radio stopped producing formal fifteen-minute to hourly programs, a new format developed. "Top 40" was based on a continuous rotation of short pop songs presented by a "disc jockey." Famous disc jockeys in the era included Alan Freed, Dick Clark, Don Imus and Wolfman Jack. Top 40 playlists were theoretically based on record sales; however, record companies began to bribe disc jockeys to play selected artists, in a controversy that was called "payola". In the 1950s, American television networks introduced broadcasts in color. The Federal Communications Commission approved the world's first monochrome-compatible color television standard in December 1953. The first network colorcast followed on January 1, 1954, with NBC transmitting the annual Tournament of Roses Parade in Pasadena, California to over 20 stations across the country. An educational television network, National Educational Television (NET), predecessor to PBS, was founded. Shortwave broadcasting played an important part of fighting the cold war with Voice of America and the BBC World Service, Radio Free Europe and Radio Liberty transmitting through the "Iron Curtain", and Radio Moscow and others broadcasting back, as well as "jamming" (transmitting to cause intentional interference) the Western stations in the Soviet bloc.

1970s, 1980s, and 1990s

Australia

The Australian Broadcasting Corporation logo, first introduced in 1975 and based on the Lissajous curve.

After much procrastination on the part of various federal governments, FM broadcasting was eventually introduced in 1975. (There had been official experiments with FM broadcasting as far back as 1948.) Only a handful of radio stations were given new licences during the 1940s, 50s & 60s but, since 1975, many hundreds of new broadcasting licences have been issued on both the FM and AM bands. In the latter case, this was made possible by having 9 kHz between stations, rather 10 kHz breaks, as per the Geneva Frequency Plan. The installation of directional aerials also encouraged more AM stations. The type of station given FM licences reflects the policies and philosiphies of the various Australian governments. Initially, only the ABC and community radio stations were granted FM licences. However, after a change of government, commercial stations were permitted on the band, as from 1980. At first, one or two brand new stations were permitted in each major market. However, in 1990, one or two existing AM stations in each major market were given FM licences; the stations being chosen by an auction system. Apart from an initial settling-in period for those few stations transferred from AM to FM, there has been no simulcasting between AM and FM stations. In major cities, a number of brand new FM licences were issued in the 1990s and 2000s. All rural regions which traditionally had only one commercial station now have at least one AM and one FM commercial station. In many cases, the owner of the original station now has at least two outlets. The number of regional transmitters for the ABC's five networks also increased dramatically during this era.

Europe

Commercial radio (re-)legalisation in most European countries occurred in this era, starting with United Kingdom in 1973 and ending with Austria in 1995. In 1987, stations in the European Broadcasting Union began offering Radio Data System (RDS), which provides written text information about programs that were being broadcast, as well as traffic alerts, accurate time, and other teletext services.

Sri Lanka

The Government of Sri Lanka opened up the market in the late 1970s and 1980s allowing private companies to set up radio and television stations. Sri Lanka's public services broadcasters are the Sri Lanka Broadcasting Corporation (SLBC), Independent Television Net Work (ITN) and the affiliated radio station called Lak-handa. They had stiff competition on their hands with the private sector. Broadcasting in Sri Lanka went through a transformation resulting in private broadcasting institutions being set up on the island among them Telshan Network (Pvt) Ltd (TNL), Maharaja Television – TV, Sirasa TV and Shakthi TV, and EAP Network (Pvt) Ltd – known as Swarnawahini – these private channels all have radio stations as well. The 1990s saw a new generation of radio stations being established in Sri Lanka among them the 'Hiru' radio station. In the 1980s public service broadcasters like the Sri Lanka Broadcasting Corporation set up their own FM arm. Sri Lanka celebrated 80 years of broadcasting in December 2005. In January 2007 the Sri Lanka Broadcasting Corporation celebrated 40 years as a public corporation.

United Kingdom

A new pirate station, Swiss-owned Radio Nordsee International, broadcast to Britain and the Netherlands from 1970 until outlawed by Dutch legislation in 1974 (which meant it could no longer be supplied from the European mainland). The English service was heavily jammed by both Labour and Conservative Governments in 1970 amid suggestions that the ship was actually being used for espionage. Radio Caroline returned in 1972 and continued until its ship sank in 1980 (the crew were rescued). A Belgian station, Radio Atlantis, operated an English service for a few months before the Dutch act came into force in 1974. Land-based commercial radio finally came on air in 1973 with London's LBC and Capital Radio. Channel 4 television started in November 1982. Britain's UHF system was originally designed to carry only four networks. Pirate radio enjoyed another brief resurgence with a literal re-launch of Radio Caroline in 1983, and the arrival of American-owned Laser 558 in 1985. Both stations were harassed by the British authorities; Laser closed in 1987 and Caroline in 1989, since then it has pursued legal methods of broadcasting, such as temporary FM licences and satellite. Two rival satellite television systems came on the air at the end of the 1980s: Sky Television and British Satellite Broadcasting. Huge losses forced a rapid merger, although in many respects it was a takeover of BSB (Britain's official, Government-sanctioned satellite company) by Sky. Radio Luxembourg launched a 24-hour English channel on satellite, but closed its AM service in 1989 and its satellite service in 1991.

The Broadcasting Act 1990 in UK law marked the establishment of two licensing authorities – the Radio Authority and the Independent Television Commission – to facilitate the licensing of non-BBC broadcast services, especially short-term broadcasts. Channel 5 went on the air on March 30, 1997, using "spare" frequencies between the existing channels.

United States

The rise of FM changed the listening habits of younger Americans. Many stations such as WNEW-FM in New York City began to play whole sides of record albums, as opposed to the "Top 40" model of two decades earlier. In the 1980s, the Federal Communications Commission, under Reagan Administration and Congressional pressure, changed the rules limiting the number of radio and television stations a business entity could own in one metropolitan area. This deregulation led to several groups, such as Infinity Broadcasting and Clear Channel to buy many stations in major cities. The cost of these stations' purchases led to a conservative approach to broadcasting, including limited playlists and avoiding controversial subjects to not offend listeners, and increased commercials to increase revenue. AM radio declined throughout the 1970s and 1980s due to various reasons including: Lower cost of FM receivers, narrow AM audio bandwidth, and poor sound in the AM section of automobile receivers (to combat the crowding of stations in the AM band and a "loudness war" conducted by AM broadcasters), and increased radio noise in homes caused by fluorescent lighting and introduction of electronic devices in homes. AM radio's decline flattened out in the mid-1990s due to the introduction of niche formats and over commercialization of many FM stations.

The 2000s

The 2000s saw the introduction of digital radio and direct broadcasting by satellite (DBS) in the USA. Digital radio services, except in the United States, were allocated a new frequency band in the range of 1,400 MHz. Regular shortwave broadcasts using Digital Radio Mondiale (DRM), a digital broadcasting scheme for short and medium wave broadcasts have begun. This system makes the normally scratchy international broadcasts clear and nearly FM quality, and much lower transmitter power. This is much better to listen to and has more languages.

The "Kerbango Internet Radio" was the first stand-alone product that let users listen to Internet radio without a computer.

Australia

In Australia, from August 2009, digital radio was phased in by geographical region. Today, the ABC, SBS, commercial and community radio stations operate on the AM and FM bands. Most stations are available on the internet and most also have digital outlets. By 2007, there were 261 commercial stations in Australia. The ABC currently has five AM/FM networks and is in the process of establishing a series of supplementary music stations that are only available on digital radios and digital television sets. SBS provides non-English language programs over its two networks, as do a number of community radio stations.

Canada

In Canada, the Canadian Radio-television and Telecommunications Commission plans to move all Canadian broadcasting to the digital band and close all mediumwave and FM stations.

Europe

European stations have begun digital broadcasting (DAB). Digital radios began to be sold in the United Kingdom in 1998.

Sri Lanka

In Sri Lanka in 2005 when Sri Lanka celebrated 80 years in Broadcasting, the former Director-General of the Sri Lanka Broadcasting Corporation, Eric Fernando called for the station to take full advantage of the digital age – this included looking at the archives of Radio Ceylon. Ivan Corea asked the President of Sri Lanka, Mahinda Rajapakse to invest in the future of the SLBC.

United States

In the United States, this band was deemed to be vital to national defense, so an alternate band in the range of 2,300 MHz was introduced for satellite broadcasting. Two American companies, XM and Sirius, introduced DBS systems, which are funded by direct subscription, as in cable television. The XM and Sirius systems provide approximately 100 channels each, in exchange for monthly payments. In addition, a consortium of companies received FCC approval for In-Band On-Channel digital broadcasts in the United States, which use the existing mediumwave and FM bands to provide CD-quality sound. However, early IBOC tests showed interference problems with adjacent channels, which has slowed adoption of the system.

4

Fundamental Concepts of Broadcasting

The chapter strategically encompasses and incorporates the major components and key concepts of broadcasting, providing a complete understanding. Some of the topics discussed in this chapter include radio broadcasting, digital audio broadcasting, commercial broadcasting, public broadcasting, remote broadcast, microbroadcasting, broadcast syndication, simulcast, broadcast signal intrusion etc.

Radio Broadcasting

Radio broadcasting is a unidirectional wireless transmission over radio waves intended to reach a wide audience. Stations can be linked in radio networks to broadcast a common radio format, either in broadcast syndication or simulcast or both. Audio broadcasting also can be done via cable radio, local wire television networks, satellite radio, and internet radio via streaming media on the Internet. The signal types can be either analog audio or digital audio.

Long wave radio broadcasting station, Motala, Sweden

Slovak Radio Building, Bratislava, Slovakia (architects: Štefan Svetko, Štefan Ďurkovič and Barnabáš Kissling, 1967-1983)

Broadcasting tower in Trondheim, Norway

History

The earliest radio stations were simply radiotelegraphy systems and did not carry audio. For audio broadcasts to be possible electronic detection and amplification devices had to be incorporated.

The thermionic valve was invented in 1904 by the English physicist John Ambrose Fleming. He developed a device he called an "oscillation valve" (because it passes current in only one direction). The heated filament, or cathode, was capable of thermionic emission of electrons that would flow to the *plate* (or *anode*) when it was at a higher voltage. Electrons, however, could not pass in the reverse direction because the plate was not heated and thus not capable of thermionic emission of electrons. Later known as the Fleming valve, it could be used as a rectifier of alternating current and as a radio wave detector. This greatly improved the crystal set which rectified the radio signal using an early solid-state diode based on a crystal and a so-called cat's whisker. However, what was still required was an amplifier.

The triode (mercury-vapor filled with a control grid) was patented on March 4, 1906 by the Austrian Robert von Lieben independent from that, on October 25, 1906 Lee De Forest patented his three-element Audion. It wasn't put to practical use until 1912, when its amplifying ability became recognized by researchers.

By about 1920, valve technology had matured to the point where radio broadcasting was quickly becoming viable. However, an early audio transmission that could be termed a *broadcast* may have occurred on Christmas Eve in 1906 by Reginald Fessenden, although this is disputed. While many early experimenters attempted to create systems similar to radiotelephone devices by which only two parties were meant to communicate, there were others who intended to transmit to larger audiences. Charles Herrold started broadcasting in California in 1909 and was carrying audio by the next year. (Herrold's station eventually became KCBS).

In The Hague, the Netherlands, PCGG started broadcasting on November 6, 1919, making it, arguably the first commercial broadcasting station. In 1916, Frank Conrad, an electrical engineer employed at the Westinghouse Electric Corporation, began broadcasting from his Wilkinsburg, Pennsylvania garage with the call letters 8XK. Later, the station was moved to the top of the Westinghouse factory building in East Pittsburgh, Pennsylvania. Westinghouse relaunched the station as KDKA on November 2, 1920, as the first commercially licensed radio station in America. The commercial broadcasting designation came from the type of broadcast license; advertisements did not air until years later. The first licensed broadcast in the United States came from KDKA itself: the results of the Harding/Cox Presidential Election. The Montreal station that became CFCF began broadcast programming on May 20, 1920, and the Detroit station that became WWJ began program broadcasts beginning on August 20, 1920, although neither held a license at the time.

In 1920 wireless broadcasts for entertainment began in the UK from the Marconi Research Centre 2MT at Writtle near Chelmsford, England. A famous broadcast from Marconi's New Street Works factory in Chelmsford was made by the famous soprano Dame Nellie Melba on 15 June 1920, where she sang two arias and her famous trill. She was the first artist of international renown to participate in direct radio broadcasts. The 2MT station began to broadcast regular entertainment in 1922. The BBC was amalgamated in 1922 and received a Royal Charter in 1926, making it the first national broadcaster in the world.

Radio Argentina began regularly scheduled transmissions from the Teatro Coliseo in Buenos Aires on August 27, 1920, making its own priority claim. The station got its license on November 19, 1923. The delay was due to the lack of official Argentine licensing procedures before that date. This station continued regular broadcasting of entertainment and cultural fare for several decades.

Radio in education soon followed and colleges across the U.S. began adding radio broadcasting courses to their curricula. Curry College in Milton, Massachusetts introduced one of the first broadcasting majors in 1932 when the college teamed up with WLOE in Boston to have students broadcast programs.

Types

Broadcasting by radio takes several forms. These include AM and FM stations. There are several subtypes, namely commercial broadcasting, non-commercial educational (NCE) public broadcasting and non-profit varieties as well as community radio, student-run campus radio stations and hospital radio stations can be found throughout the world.

Many stations broadcast on shortwave bands using AM technology that can be received over thousands of miles (especially at night). For example, the BBC, VOA, VOR, and Deutsche Welle have transmitted via shortwave to Africa and Asia. These broadcasts are very sensitive to atmospheric conditions and solar activity.

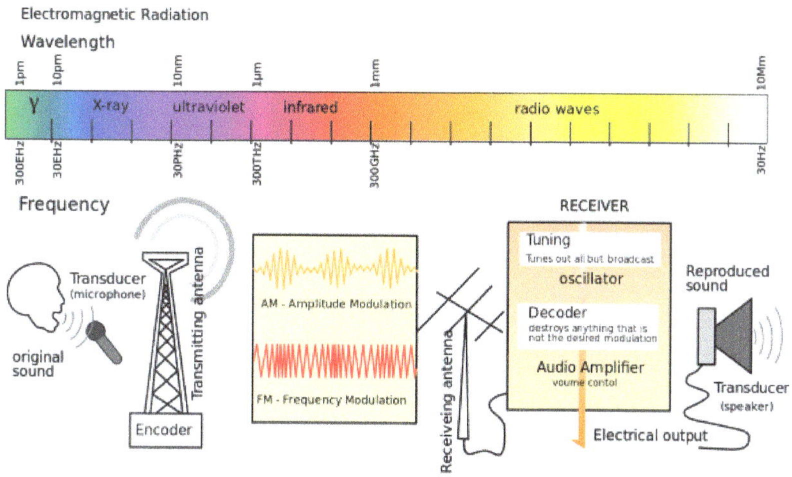

Transmission diagram of sound broadcasting (AM and FM)

Nielsen Audio, formerly known as Arbitron, the United States-based company that reports on radio audiences, defines a "radio station" as a government-licensed AM or FM station; an HD Radio (primary or multicast) station; an internet stream of an existing government-licensed station; one of the satellite radio channels from XM Satellite Radio or Sirius Satellite Radio; or, potentially, a station that is not government licensed.

Shortwave

See shortwave for the differences between shortwave, medium wave and long wave spectra. Shortwave is used largely for national broadcasters, international propaganda, or religious broadcasting organizations.

AM

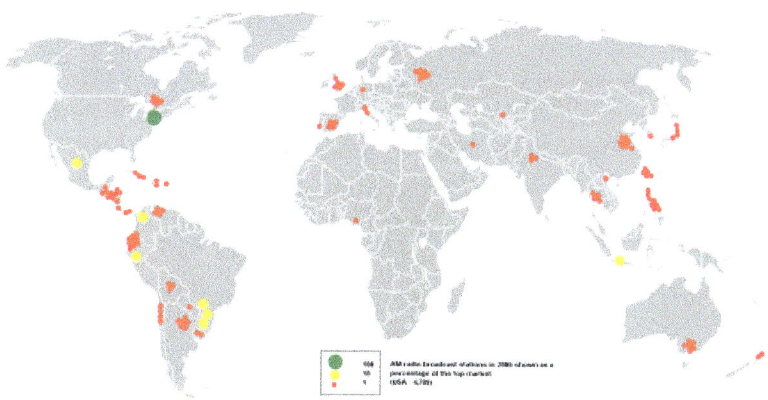

AM broadcasting stations in 2006

AM stations were the earliest broadcasting stations to be developed. AM refers to amplitude modulation, a mode of broadcasting radio waves by varying the amplitude of the carrier signal in response to the amplitude of the signal to be transmitted.

The medium-wave band is used worldwide for AM broadcasting. Europe also uses the long wave band. In response to the growing popularity of FM stereo radio stations in the late 1980s and early 1990s, some North American stations began broadcasting in AM stereo, though this never gained popularity, and very few receivers were ever sold.

One of the advantages of AM is that its signal can be detected (turned into sound) with simple equipment. If a signal is strong enough, not even a power source is needed; building an unpowered crystal radio receiver was a common childhood project in the early decades of AM broadcasting.

AM broadcasts occur on North American airwaves in the medium wave frequency range of 530 to 1700 kHz (known as the "standard broadcast band"). The band was expanded in the 1990s by adding nine channels from 1620 to 1700 kHz. Channels are spaced every 10 kHz in the Americas, and generally every 9 kHz everywhere else.

The signal is subject to interference from electrical storms (lightning) and other electromagnetic interference (EMI).

AM transmissions cannot be ionospherically propagated during the day due to strong absorption in the D-layer of the ionosphere. In a crowded channel environment this means that the power of regional channels which share a frequency must be reduced at night or directionally beamed in order to avoid interference, which reduces the potential nighttime audience. Some stations have frequencies unshared with other stations in North America; these are called clear-channel stations. Many of them can be heard across much of the country at night. During the night, absorption largely disappears and permits signals to travel to much more distant locations via ionospheric reflections. However, fading of the signal can be severe at night.

AM radio transmitters can transmit audio frequencies up to 15 kHz (now limited to 10 kHz in the US due to FCC rules designed to reduce interference), but most receivers are only capable of reproducing frequencies up to 5 kHz or less. At the time that AM broadcasting began in the 1920s, this provided adequate fidelity for existing microphones, 78 rpm recordings, and loudspeakers. The fidelity of sound equipment subsequently improved considerably, but the receivers did not. Reducing the bandwidth of the receivers reduces the cost of manufacturing and makes them less prone to interference. AM stations are never assigned adjacent channels in the same service area. This prevents the sideband power generated by two stations from interfering with each other. Bob Carver created an AM stereo tuner employing notch filtering that demonstrated that an AM broadcast can meet or exceed the 15 kHz baseband bandwidth allotted to FM stations without objectionable interference. After several years, the tuner was discontinued. Bob Carver had left the company and the Carver Corporation later cut the number of models produced before discontinuing production completely.

FM

FM refers to frequency modulation, and occurs on VHF airwaves in the frequency range of 88 to 108 MHz everywhere except Japan and Russia. Russia, like the former Soviet Union, uses 65.9 to

74 MHz frequencies in addition to the world standard. Japan uses the 76 to 90 MHz frequency band.

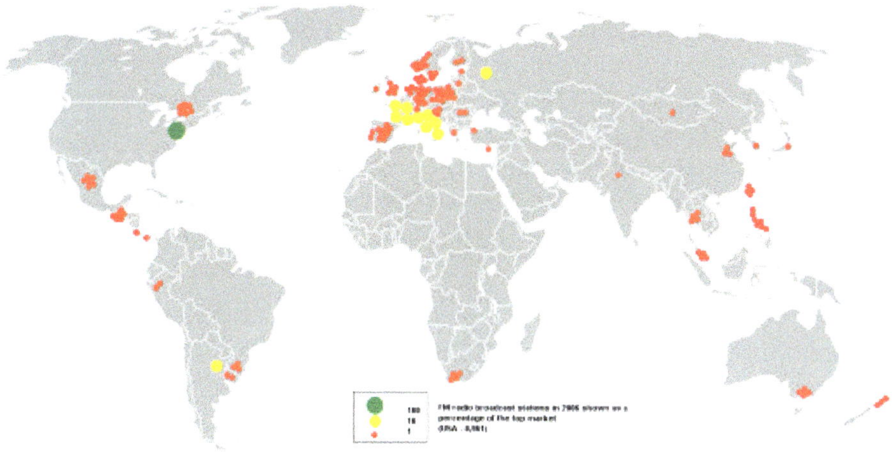

FM radio broadcast stations in 2006

Edwin Howard Armstrong invented FM radio to overcome the problem of radio-frequency interference (RFI), which plagued AM radio reception. At the same time, greater fidelity was made possible by spacing stations further apart in the radio frequency spectrum. Instead of 10 kHz apart, as on the AM band in the US, FM channels are 200 kHz (0.2 MHz) apart. In other countries greater spacing is sometimes mandatory, such as in New Zealand, which uses 700 kHz spacing (previously 800 kHz). The improved fidelity made available was far in advance of the audio equipment of the 1940s, but wide interchannel spacing was chosen to take advantage of the noise-suppressing feature of wideband FM.

Bandwidth of 200 kHz is not needed to accommodate an audio signal — 20 kHz to 30 kHz is all that is necessary for a narrowband FM signal. The 200 kHz bandwidth allowed room for ±75 kHz signal deviation from the assigned frequency, plus guard bands to reduce or eliminate adjacent channel interference. The larger bandwidth allows for broadcasting a 15 kHz bandwidth audio signal plus a 38 kHz stereo "subcarrier"—a piggyback signal that rides on the main signal. Additional unused capacity is used by some broadcasters to transmit utility functions such as background music for public areas, GPS auxiliary signals, or financial market data.

The AM radio problem of interference at night was addressed in a different way. At the time FM was set up, the available frequencies were far higher in the spectrum than those used for AM radio - by a factor of approximately 100. Using these frequencies meant that even at far higher power, the range of a given FM signal was much shorter; thus its market was more local than for AM radio. The reception range at night is the same as in the daytime. All FM broadcast transmissions are line-of-sight, and ionospheric bounce is not viable. The much larger bandwidths, compared to AM and SSB, are more susceptible to phase dispersion. Propagation speeds (celerities) are fastest in the ionosphere at the lowest sideband frequency. The celerity difference between the highest and lowest sidebands is quite apparent to the listener. Such distortion occurs up to frequencies of approximately 50 MHz. Higher frequencies do not reflect from the ionosphere, nor from storm clouds. Moon reflections have been used in some experiments, but require impractical power levels.

The original FM radio service in the U.S. was the Yankee Network, located in New England. Regular FM broadcasting began in 1939, but did not pose a significant threat to the AM broadcasting industry. It required purchase of a special receiver. The frequencies used, 42 to 50 MHz, were not those used today. The change to the current frequencies, 88 to 108 MHz, began after the end of World War II, and was to some extent imposed by AM broadcasters as an attempt to cripple what was by now realized to be a potentially serious threat.

FM radio on the new band had to begin from the ground floor. As a commercial venture it remained a little-used audio enthusiasts' medium until the 1960s. The more prosperous AM stations, or their owners, acquired FM licenses and often broadcast the same programming on the FM station as on the AM station ("simulcasting"). The FCC limited this practice in the 1960s. By the 1980s, since almost all new radios included both AM and FM tuners, FM became the dominant medium, especially in cities. Because of its greater range, AM remained more common in rural environments.

Pirate Radio

Pirate radio is illegal or non-regulated radio transmission. It is most commonly used to describe illegal broadcasting for entertainment or political purposes. Sometimes it is used for illegal two-way radio operation. Its history can be traced back to the unlicensed nature of the transmission, but historically there has been occasional use of sea vessels—fitting the most common perception of a pirate—as broadcasting bases. Rules and regulations vary largely from country to country, but often the term pirate radio generally describes the unlicensed broadcast of FM radio, AM radio, or short wave signals over a wide range. In some places radio stations are legal where the signal is transmitted, but illegal where the signals are received—especially when the signals cross a national boundary. In other cases, a broadcast may be considered "pirate" due to the type of content, its transmission format, or the transmitting power (wattage) of the station, even if the transmission is not technically illegal (such as a web cast or an amateur radio transmission). Pirate radio stations are sometimes referred to as bootleg radio or clandestine stations.

Terrestrial Digital Radio

Digital radio broadcasting has emerged, first in Europe (the UK in 1995 and Germany in 1999), and later in the United States, France, the Netherlands, South Africa and many other countries worldwide. The most simple system is named DAB Digital Radio, for Digital Audio Broadcasting, and uses the public domain EUREKA 147 (Band III) system. DAB is used mainly in the UK and South Africa. Germany and the Netherlands use the DAB and DAB+ systems, and France uses the L-Band system of DAB Digital Radio.

In the United States, digital radio isn't used in the same way as Europe and South Africa. Instead, the IBOC system is named HD Radio and owned by a consortium of private companies that is called iBiquity. An international non-profit consortium Digital Radio Mondiale (DRM), has introduced the public domain DRM system.

Satellite

Satellite radio broadcasters are slowly emerging, but the enormous entry costs of space-based satellite transmitters, and restrictions on available radio spectrum licenses has restricted growth of this market. In the USA and Canada, just two services, XM Satellite Radio and Sirius Satellite Radio exist. Both XM and Sirius are owned by Sirius XM Radio, which was formed by the merger of XM and Sirius on July 29, 2008, whereas in Canada, XM Radio Canada and Sirius Canada remained separate companies until 2010. Worldspace in Africa and Asia, and MobaHO! in Japan and the ROK were two unsuccessful satellite radio operators which have gone out of business.

Program Formats

Radio program formats differ by country, regulation and markets. For instance, the U.S. Federal Communications Commission designates the 88–92 megahertz band in the U.S. for non-profit or educational programming, with advertising prohibited.

In addition, formats change in popularity as time passes and technology improves. Early radio equipment only allowed program material to be broadcast in real time, known as *live* broadcasting. As technology for sound recording improved, an increasing proportion of broadcast programming used pre-recorded material. A current trend is the automation of radio stations. Some stations now operate without direct human intervention by using entirely pre-recorded material sequenced by computer control.

Digital Audio Broadcasting

Digital audio broadcasting (DAB) is a digital radio technology for broadcasting radio stations, used in several countries across Europe and Asia Pacific.

The DAB standard was initiated as a European research project in the 1980s. The Norwegian Broadcasting Corporation (NRK) launched the first DAB channel in the world on 1 June 1995 (NRK Klassisk), and the BBC and Swedish Radio (SR) launched their first DAB digital radio broadcasts in September 1995. DAB receivers have been available in many countries since the end of the 1990s.

DAB may offer more radio programmes over a specific spectrum than analogue FM radio. DAB is more robust with regard to noise and multipath fading for mobile listening, since DAB reception quality first degrades rapidly when the signal strength falls below a critical threshold, whereas FM reception quality degrades slowly with the decreasing signal.

Audio quality varies depending on the bitrate used and audio material. Most stations use a bit rate of 128 kbit/s or less with the MP2 audio codec, which requires 160 kbit/s to achieve perceived FM quality. 128 kbit/s gives better dynamic range or signal-to-noise ratio than FM radio, but a more smeared stereo image, and an upper cut-off frequency of 14 kHz, corresponding to 15 kHz of FM radio. However, "CD sound quality" with MP2 is possible "with 256...192 kbps".

An upgraded version of the system was released in February 2007, which is called DAB+. DAB is not forward compatible with DAB+, which means that DAB-only receivers are not able to receive DAB+ broadcasts. However, broadcasters can mix DAB and DAB+ programs inside the same transmission and so make a progressive transition to DAB+. DAB+ is approximately twice as efficient as DAB due to the adoption of the AAC+ audio codec, and DAB+ can provide high quality audio with bit rates as low as 64 kbit/s. Reception quality is also more robust on DAB+ than on DAB due to the addition of Reed-Solomon error correction coding.

In spectrum management, the bands that are allocated for public DAB services, are abbreviated with T-DAB, where the "T" stands for terrestrial.

More than 30 countries provide DAB transmissions, and several countries, such as Norway, Australia, Italy, Malta, Switzerland, The Netherlands and Germany, are transmitting DAB+ stations.

In many countries it is expected that DAB will gradually replace FM radio. Norway was the first country to announce national FM radio switch-off starting from 2017.

History

DAB has been under development since 1981 at the *Institut für Rundfunktechnik* (IRT). In 1985 the first DAB demonstrations were held at the WARC-ORB in Geneva and in 1988 the first DAB transmissions were made in Germany. Later DAB was developed as a research project for the European Union (EUREKA), which started in 1987 on initiative by a consortium formed in 1986. The MPEG-1 Audio Layer II ("MP2") codec was created as part of the EU147 project. DAB was the first standard based on orthogonal frequency division multiplexing (OFDM) modulation technique, which since then has become one of the most popular transmission schemes for modern wideband digital communication systems.

- Countries with regular services
- Countries with trials / tests
- Countries with interest
- DAB no longer used

(2012)

A choice of audio codec, modulation and error-correction coding schemes and first trial broadcasts were made in 1990. Public demonstrations were made in 1993 in the United Kingdom. The protocol specification was finalized in 1993 and adopted by the ITU-R standardization body in 1994, the European community in 1995 and by ETSI in 1997. Pilot broadcasts were launched in several countries in 1995.

The UK was the first country to receive a wide range of radio stations via DAB. Commercial DAB receivers began to be sold in 1999 and over 50 commercial and BBC services were available in London by 2001.

The standard was coordinated by the European DAB forum, formed in 1995 and reconstituted to the World DAB Forum in 1997, which represents more than 30 countries. In 2006 the World DAB Forum became the World DMB Forum which now presides over both the DAB and DMB standard.

In October 2005, the World DMB Forum instructed its Technical Committee to carry out the work needed to adopt the AAC+ audio codec and stronger error correction coding. This work led to the launch of the new DAB+ system.

By 2006, 500 million people worldwide were in the coverage area of DAB broadcasts, although by this time sales had only taken off in the United Kingdom and Denmark. In 2006 there were approximately 1,000 DAB stations in operation worldwide.

Technology

Bands and Modes

DAB uses a wide-bandwidth broadcast technology and typically spectra have been allocated for it in Band III (174–240 MHz) and L band (1,452–1,492 MHz), although the scheme allows for operation almost anywhere above 30 MHz. The US military has reserved L-Band in the USA only, blocking its use for other purposes in America, and the United States has reached an agreement with Canada to restrict L-Band DAB to terrestrial broadcast to avoid interference.

DAB has a number of country specific transmission modes (I, II, III and IV). For worldwide operation a receiver must support all 4 modes:

- Mode I for Band III, Earth
- Mode II for L-Band, Earth and satellite
- Mode III for frequencies below 3 GHz, Earth and satellite
- Mode IV for L-Band, Earth and satellite

Protocol Stack

From an OSI model protocol stack viewpoint, the technologies used on DAB inhabit the following layers: the audio codec inhabits the presentation layer. Below that is the data link layer, in charge of statistical time division multiplexing and frame synchronization. Finally, the physical layer contains the error-correction coding, OFDM modulation, and dealing with the over-the-air transmission and reception of data. Some aspects of these are described below.

Audio Codec

The older version of DAB that is being used in Denmark*, Ireland*, Norway*, Switzerland* and the UK, uses the MPEG-1 Audio Layer 2 audio codec, which is also known as *MP2* due to computer files using those characters for their file extension. Denmark, Ireland, Norway, and Switzerland also use DAB+.

The new DAB+ standard has adopted the HE-AAC version 2 audio codec, commonly known as 'AAC+' or 'aacPlus'. AAC+ is approximately three-times more efficient than MP2, which means that broadcasters using DAB+ will be able to provide far higher audio quality or far more stations than they can on DAB, or, as is most likely, a combination of both higher audio quality and more stations will be provided.

One of the most important decisions regarding the design of a digital radio system is the choice of which audio codec to use, because the efficiency of the audio codec determines how many radio stations can be carried on a multiplex at a given level of audio quality. The capacity of a DAB multiplex is fixed, so the more efficient the audio codec is, the more stations can be carried, and vice versa. Similarly, for a fixed bit-rate level, the more efficient the audio codec is the higher the audio quality will be.

Error-correction Coding

Error-correction coding (ECC) is an important technology for a digital communication system because it determines how robust the reception will be for a given signal strength – stronger ECC will provide more robust reception than a weaker form.

The old version of DAB uses punctured convolutional coding for its ECC. The coding scheme uses unequal error protection (UEP), which means that parts of the audio bit-stream that are more susceptible to errors causing audible disturbances are provided with more protection (i.e. a lower code rate) and vice versa. However, the UEP scheme used on DAB results in there being a grey area in between the user experiencing good reception quality and no reception at all, as opposed to the situation with most other wireless digital communication systems that have a sharp "digital cliff", where the signal rapidly becomes unusable if the signal strength drops below a certain threshold. When DAB listeners receive a signal in this intermediate strength area they experience a "burbling" sound which interrupts the playback of the audio.

The new DAB+ standard has incorporated Reed-Solomon ECC as an "inner layer" of coding that is placed around the byte interleaved audio frame but inside the "outer layer" of convolutional coding used by the older DAB system, although on DAB+ the convolutional coding uses equal error protection (EEP) rather than UEP since each bit is equally important in DAB+. This combination of Reed-Solomon coding as the inner layer of coding, followed by an outer layer of convolutional coding – so-called "concatenated coding" – became a popular ECC scheme in the 1990s, and NASA adopted it for its deep-space missions. One slight difference between the concatenated coding used by the DAB+ system and that used on most other systems is that it uses a rectangular byte interleaver rather than Forney interleaving in order to provide a greater interleaver depth, which increases the distance over which error bursts will be spread out in the bit-stream, which in turn will allow the Reed-Solomon error decoder to correct a higher proportion of errors.

The ECC used on DAB+ is far stronger than is used on DAB, which, with all else being equal (i.e. if the transmission powers remained the same), would translate into people who currently experience reception difficulties on DAB receiving a much more robust signal with DAB+ transmissions. It also has a far steeper "digital cliff", and listening tests have shown that people prefer this when the signal strength is low compared to the shallower digital cliff on DAB.

Modulation

Immunity to fading and inter-symbol interference (caused by multipath propagation) is achieved without equalization by means of the OFDM and DQPSK modulation techniques.

Using values for the most commonly used transmission mode on DAB, Transmission Mode I (TM I), the OFDM modulation consists of 1,536 subcarriers that are transmitted in parallel. The useful part of the OFDM symbol period is 1 millisecond, which results in the OFDM subcarriers each having a bandwidth of 1 kHz due to the inverse relationship between these two parameters, and the overall OFDM channel bandwidth is 1,537 kHz. The OFDM guard interval for TM I is 246 microseconds, which means that the overall OFDM symbol duration is 1.246 milliseconds. The guard interval duration also determines the maximum separation between transmitters that are part of the same single-frequency network (SFN), which is approximately 74 km for TM I.

Single-frequency Networks

OFDM allows the use of single-frequency networks (SFN), which means that a network of transmitters can provide coverage to a large area – up to the size of a country – where all transmitters use the same transmission frequency. Transmitters that are part of an SFN need to be very accurately synchronised with other transmitters in the network, which requires the transmitters to use very accurate clocks.

When a receiver receives a signal that has been transmitted from the different transmitters that are part of an SFN, the signals from the different transmitters will typically have different delays, but to OFDM they will appear to simply be different multipaths of the same signal. Reception difficulties can arise, however, when the relative delay of multipaths exceeds the OFDM guard interval duration, and there are frequent reports of reception difficulties due to this issue when there is a *lift*, such as when there's high pressure, due to signals travelling farther than usual, and thus the signals are likely to arrive with a relative delay that is greater than the OFDM guard interval.

Low power *gap-filler* transmitters can be added to an SFN as and when desired in order to improve reception quality, although the way SFNs have been implemented in the UK up to now they have tended to consist of higher power transmitters being installed at main transmitter sites in order to keep costs down.

Bit Rates

An ensemble has a maximum bit rate that can be carried, but this depends on which error protection level is used. However, all DAB multiplexes can carry a total of 864 "capacity units". The number of capacity units, or CU, that a certain bit-rate level requires depends on the amount of error correction added to the transmission, as described above. In the UK, most services transmit using 'protection level three', which provides an average ECC code rate of approximately ½, equating to a maximum bit rate per multiplex of 1,184 kbit/s.

Services and Ensembles

Various different services are embedded into one ensemble (which is also typically called a multiplex). These services can include:

- Primary services, like main radio stations
- Secondary services, like additional sports commentaries
- Data services
    - Electronic Programme Guide (EPG)
    - Collections of HTML pages and digital images (Known as 'Broadcast Web Sites')
    - Slideshows, which may be synchronised with audio broadcasts. For example, a police appeal could be broadcast with the e-fit of a suspect or CCTV footage.
    - Video
    - Java Platform Applications
    - IP tunnelling
    - Other raw data

DAB+

The term DAB most commonly refers both to a specific DAB standard using the MP2 audio codec, but can sometimes refer to a whole family of DAB related standards, such as DAB+, DMB and DAB-IP.

DAB+

WorldDAB, the organisation in charge of the DAB standards, announced DAB+, a major upgrade to the DAB standard in 2006, when the HE-AAC v2 audio codec (also known as eAAC+) was adopted. The new standard, which is called DAB+, has also adopted the MPEG Surround audio format and stronger error correction coding in the form of Reed-Solomon coding. DAB+ has been standardised as European Telecommunications Standards Institute (ETSI) TS 102 563.

As DAB is not forward compatible with DAB+, older DAB receivers can not receive DAB+ broadcasts. However, DAB receivers that will be able to receive the new DAB+ standard via a firmware upgrade went on sale in July 2007. If a receiver is DAB+ compatible, there will be a sign on the product packaging.

DAB+ broadcasts have launched in several countries like Australia, Czech Republic, Denmark, Germany, Hong Kong, Italy, Malta, Norway, Poland, Switzerland, the United Kingdom and the Netherlands. Malta was the first country to launch DAB+ in Europe. Several other countries are also expected to launch DAB+ broadcasts over the next few years, such as Austria, Hungary and Asian countries, such as Thailand, Vietnam and Indonesia. South Africa began a DAB+ technical pilot in November 2014 on channel 13F in Band 3. If DAB+ stations launch in established DAB countries, they can transmit alongside existing DAB stations that use the older MPEG-1 Audio Layer II audio format, and most existing DAB stations are expected to continue broadcasting until the vast majority of receivers support DAB+.

Ofcom in the UK has published a consultation with the intention to set up a new multiplex containing a mix of DAB and DAB+ services, with the intention of moving services to this format in the long term. In February 2016 3 DAB+ stations launched in the UK as part of a new national network.

DMB

Digital multimedia broadcasting (DMB) and DAB-IP are suitable for mobile radio and TV both because they support MPEG 4 AVC and WMV9 respectively as video codecs. However, a DMB video subchannel can easily be added to any DAB transmission, as it was designed to be carried on a DAB subchannel. DMB broadcasts in Korea carry conventional MPEG 1 Layer II DAB audio services alongside their DMB video services.

Norway, South Korea and France are countries currently[*when?*] broadcasting DMB.

Countries Using DAB

More than 30 countries provide DAB, DAB+ and/or DMB broadcasts, either as a permanent technology or as test transmissions.

DAB and AM/FM Compared

Traditionally radio programmes were broadcast on different frequencies via AM and FM, and the radio had to be tuned into each frequency, as needed. This used up a comparatively large amount of spectrum for a relatively small number of stations, limiting listening choice. DAB is a digital radio broadcasting system that through the application of multiplexing and compression combines multiple audio streams onto a relatively narrow band centred on a single broadcast frequency called a DAB ensemble.

Within an overall target bit rate for the DAB ensemble, individual stations can be allocated different bit rates. The number of channels within a DAB ensemble can be increased by lowering average bit rates, but at the expense of the quality of streams. Error correction under the DAB standard makes the signal more robust but reduces the total bit rate available for streams.

FM HD Radio Versus DAB

Some countries have implemented Eureka-147 digital audio broadcasting (DAB). DAB broadcasts a single station that is approximately 1,500 kilohertz wide (~1,000 kilobits per second). That station is then subdivided into multiple digital streams of between 9 and 12 programs. In contrast FM HD radio shares its digital broadcast with the traditional 200 kilohertz-wide channels, with capability of 300 kbit/s per station (pure digital mode).

The first generation DAB uses the MPEG-1 Audio Layer II (MP2) audio codec which has less efficient compression than newer codecs. The typical bitrate for DAB programs is only 128 kbit/s and as a result most radio stations on DAB have a lower sound quality than FM, prompting a number of complaints among the audiophile community. As with DAB+ or T-DMB in Europe, FM HD Radio uses a codec based upon the MPEG-4 HE-AAC standard.

HD Radio is a proprietary system from the company Ibiquity. DAB is an open standard deposited at ETSI.

Use of Frequency Spectrum and Transmitter Sites

DAB gives substantially higher spectral efficiency, measured in programmes per MHz and per transmitter site, than analogue communication. This has led to an increase in the number of stations available to listeners, especially outside of the major urban areas.

Numerical example: Analog FM requires 0.2 MHz per programme. The frequency reuse factor in most countries is approximately 15, meaning that only one out of 15 transmitter sites can use the same channel frequency without problems with co-channel interference, i.e. cross-talk. Assuming a total availability of 102 FM channels at a bandwidth of 0.2MHz over the Band II spectrum of 87.5 to 108.0 MHz, an average of 102/15 = 6.8 radio channels are possible on each transmitter site (plus lower-power local transmitters causing less interference). This results in a system spectral efficiency of 1 / 15 / (0.2 MHz) = 0.30 programmes/transmitter/MHz. DAB with 192 kbit/s codec requires 1.536 MHz * 192 kbit/s / 1,136 kbit/s = 0.26 MHz per audio programme. The frequency reuse factor for local programmes and multi-frequency broadcasting networks (MFN) is typically 4 or 5, resulting in 1 / 4 / (0.26 MHz) = 0.96 programmes/transmitter/MHz. This is 3.2 times as efficient as analog FM for local stations. For single frequency network (SFN) transmission, for example of national programmes, the channel re-use factor is 1, resulting in 1/1/0.25 MHz = 3.85 programmes/transmitter/MHz, which is 12.7 times as efficient as FM for national and regional networks.

Note the above capacity improvement may not always be achieved at the L-band frequencies, since these are more sensitive to obstacles than the FM band frequencies, and may cause shadow fading for hilly terrain and for indoor communication. The number of transmitter sites or the transmission power required for full coverage of a country may be rather high at these frequencies, to avoid the system becoming noise limited rather than limited by co-channel interference.

Sound Quality

The original objectives of converting to digital transmission were to enable higher fidelity, more stations and more resistance to noise, co-channel interference and multipath than in analogue FM radio. However, the leading countries in implementing DAB on stereo radio stations use compression to such a degree that it produces lower sound quality than that received from non-mobile FM broadcasts. This is because of the bit rate levels being too low for the MPEG Layer 2 audio codec to provide high fidelity audio quality.

The BBC Research & Development department states that at least 192 kbit/s is necessary for a high fidelity stereo broadcast :

A value of 256 kbit/s has been judged to provide a high quality stereo broadcast signal. However, a small reduction, to 224 kbit/s is often adequate, and in some cases it may be possible to accept a further reduction to 192 kbit/s, especially if redundancy in the stereo signal is exploited by a process of 'joint stereo' encoding (i.e. some sounds appearing at the centre of the stereo image need not be sent twice). At 192 kbit/s, it is relatively easy to hear imperfections in critical audio material.

—BBC R&D White Paper WHP 061 June 2003

When BBC in July 2006 reduced the bit-rate of transmission of Radio 3 from 192 kbit/s to 160 kbit/s, the resulting degradation of audio quality prompted a number of complaints to the Corporation. BBC later announced that following this testing of new equipment, it would resume the previous practice of transmitting Radio 3 at 192 kbit/s whenever there were no other demands on bandwidth.

Despite the above a survey of DAB listeners (including mobile) has shown most find DAB to have equal or better sound quality than FM.

Notwithstanding the above, BBC Radio 4 has extended the periods it broadcasts programmes with a lower bit rate (80 kbit/s) and in mono in 2012, such as the *Today* programme, rather than 128 kbit/s and in stereo. Programmes which had traditionally been broadcast on BBC Radio 4 DAB in stereo (from 1999 to 2011), can now only be heard in the evenings in mono, even though the same programmes still go out in stereo on Radio 4 FM, Digital TV and On-Line. The BBC have issued a statement stating that stereo is still their default for BBC Radio 4 DAB, however after the Olympics, this does not appear to be the case in the evenings, making FM broadcasts (in good reception areas) superior. As very few car radios are currently fitted with DAB if the BBC switch FM off as indicated later in the decade, some listeners may be forced to receive mono broadcasts in the future, a somewhat backward step.

An Audio Quality comparison of PCM, DAB, DAB+, FM and AM is available here

Benefits of DAB

Current AM and FM terrestrial broadcast technology is well established, compatible, and cheap to manufacture. Benefits of DAB over analogue systems are explained below.

Improved Features for Users

DAB radios automatically tune to all the available stations, offering a list for the user to select from.

DAB can carry "radiotext" (in DAB terminology, *Dynamic Label Segment*, or DLS) from the station giving real-time information such as song titles, music type and news or traffic updates. Advance programme guides can also be transmitted. A similar feature also exists on FM in the form of the RDS. (However, not all FM receivers allow radio stations to be stored by name.)

DAB receivers can display time of day as encoded into transmissions, so is automatically corrected when travelling between time zones and when changing to or from Daylight Saving. This is not implemented on all receivers, and some display time only when in "Standby" mode. (Similar Features on RDS: 4A Groups)

Some radios offer a pause facility on live broadcasts, caching the broadcast stream on local flash memory, although this function is limited.

More Stations

DAB is not more bandwidth efficient than analogue measured in programmes per MHz of a specific transmitter (the so-called link spectral efficiency). It is less susceptible to co-channel interference (cross talk), which makes it possible to reduce the reuse distance, i.e. use the same radio frequency channel more densely. The system spectral efficiency (the average number of radio programmes per MHz and transmitter) is a factor three more efficient than analogue FM for local radio stations, as can be seen in the above numerical example. For national and regional radio networks, the efficiency is improved by more than an order of magnitude due to the use of SFNs. In that case, adjacent transmitters use the same frequency.

In certain areas – particularly rural areas – the introduction of DAB gives radio listeners a greater choice of radio stations. For instance, in South Norway, radio listeners experienced an increase in available stations from 6 to 21 when DAB was introduced in November 2006.

Reception Quality

The DAB standard integrates features to reduce the negative consequences of multipath fading and signal noise, which afflict existing analogue systems.

Also, as DAB transmits digital audio, there is no hiss with a weak signal, which can happen on FM. However, radios in the fringe of a DAB signal, can experience a "bubbling mud" sound interrupting the audio and/or the audio cutting out altogether.

Due to sensitivity to doppler shift in combination with multipath propagation, DAB reception range (but not audio quality) is reduced when travelling speeds of more than 120 to 200 km/h, depending on carrier frequency.

Less Unlicensed ("Pirate") Station Interference

The specialised nature and cost of DAB broadcasting equipment provide barriers to unlicensed ("pirate") stations broadcasting on DAB. In cities such as London with large numbers of undocumented radio stations broadcasting on FM, this means that some stations can be reliably received via DAB in areas where they are regularly difficult or impossible to receive on FM due to undocumented radio interference.

Variable Bandwidth

Mono talk radio, news and weather channels and other non-music programs need significantly less bandwidth than a typical music radio station, which allows DAB to carry these programmes at lower bit rates, leaving more bandwidth to be used for other programs.

However, this had led to the situation where some stations are being broadcast in mono.

Transmission Costs

It is common belief that DAB is more expensive to transmit than FM. It is true that DAB uses higher frequencies than FM and therefore there is a need to compensate with more transmitters, higher radiated powers, or a combination, to achieve the same coverage. However, the last couple of years has seen significant improvement in power efficiency for DAB-transmitters.

This efficiency originates from the ability a DAB network has in broadcasting more channels per network. One network can broadcast 6–10 channels (with MPEG audio codec) or 10–16 channels (with HE AAC codec). Hence, it is thought that the replacement of FM-radios and FM-transmitters with new DAB-radios and DAB-transmitters will not cost any more as opposed to newer FM facilities.

Lower transmission costs are supported by independent network studies from Teracom (Sweden) and SSR/SRG (Switzerland). Among other things they show that DAB is as low as one-sixth of the cost of FM transmission.

Disadvantages of DAB

Reception Quality

The reception quality during the early stage of deployment of DAB was poor even for people who live well within the coverage area. The reason for this is that the old version of DAB uses weak error correction coding, so that when there are a lot of errors with the received data not enough of the errors can be corrected and a "bubbling mud" sound occurs. In some cases a complete loss of signal can happen. This situation has been improved upon in the new DAB standard (DAB+, discussed below) that uses stronger error correction coding and as additional transmitters are built. Like with DBV-T, when the signal is weak or gets interference, it will not work at all.

Audio Quality

Broadcasters have been criticized for 'squeezing in' more stations per ensemble than recommended, by:

- Minimizing the bit-rate, to the lowest level of sound-quality that listeners are willing to tolerate, such as 112 kbit/s for stereo and even 48 kbit/s for mono speech radio such as LBC 1152 and the Voice of Russia.
- Having few digital channels broadcasting in stereo.

Signal Delay

The nature of a Single-frequency network (SFN) is such that the transmitters in a network must broadcast the same signal at the same time. To achieve synchronization, the broadcaster must counter any differences in propagation time incurred by the different methods and distances involved in carrying the signal from the multiplexer to the different transmitters. This is done by applying a delay to the incoming signal at the transmitter based on a timestamp generated at the multiplexer, created taking into account the maximum likely propagation time, with a generous added margin for safety. Delays in the receiver due to digital processing (e.g. deinterleaving) add to the overall delay perceived by the listener. The signal is delayed by 2–4 seconds depending on the decoding circuitry used. This has disadvantages:

- DAB radios are out of step with live events, so the experience of listening to live commentaries on events being watched is impaired;
- Listeners using a combination of analogue (AM or FM) and DAB radios (e.g. in different rooms of a house) will hear a confusing mixture when both receivers are within earshot.

Time signals, on the contrary, are not a problem in a well-defined network with a fixed delay. The DAB multiplexer adds the proper offset to the distributed time information. The time information is also independent from the (possibly varying) audio decoding delay in receivers since the time is not embedded inside the audio frames. This means that built in clocks in receivers will be spot on.

Coverage

As DAB is at a relatively early stage of deployment, DAB coverage is poor in nearly all countries in comparison to the high population coverage provided by FM.

Exceptions include Norway, which will have 99.5% coverage by the end of 2014, and the United Kingdom, where 95% population coverage has been achieved for certain stations.

Compatibility

In 2006 tests began using the much improved HE-AAC codec for DAB+. Virtually none of the receivers made before 2008 support the new codec, however, thus making them partially obsolete once DAB+ broadcasts begin and completely obsolete once the old MPEG-1 Layer 2 stations are switched off. New receivers are both DAB and DAB+ compatible; however, the issue is exacerbated by some manufacturers disabling the DAB+ features on otherwise compatible radios to save on licensing fees when sold in countries without current DAB+ broadcasts.

Power Requirements

As DAB requires digital signal processing techniques to convert from the received digitally encoded signal to the analogue audio content, the complexity of the electronic circuitry required to do this is higher. This translates into needing more power to effect this conversion than compared to an analogue FM to audio conversion, meaning that portable receiving equipment will tend to have a shorter battery life, or require higher power (and hence more bulk). This means that they use more energy than analogue Band II VHF receivers. However, thanks to increased integration (radio-on-chip), DAB receiver is getting closer to FM one. For example, NXP semiconductor is producing both FM only and FM+DAB radio-on-chip, with similar power consumption.

As an indicator of this increased power consumption in the early days of DAB, some radio manufacturers quoted the length of time their receivers can play on a single charge. For a commonly used FM/DAB-receiver from manufacturer PURE, this is stated as: DAB 10 hours, FM 22 hours.. Currently, PURE manufacturer doesn't indicate any more power consumption difference between FM and DAB modes.

FM Radio Switch-off

Norway is the first country that has announced a complete switch-off of national FM radio stations. Switch off will start on 11 January 2017. The switch-off will not affect local and in some way regional radio stations.

Switzerland has announced its plans for a progressive digital switchover between 2020 and 2024.

Other countries like Denmark and Sweden are evaluating a switch-off by 2022.

UK is considering a progressive switch-off in the period 2017–2022.

Commercial Broadcasting

Commercial broadcasting (also called *private broadcasting*) is the broadcasting of television programs and radio programming by privately owned corporate media, as opposed to state sponsorship. It was the United States' first model of radio (and later television) during the 1920s, in contrast with the public television model in Europe during the 1930s, 1940s and 1950s which prevailed worldwide (except in the United States) until the 1980s.

Features

Advertising

Commercial broadcasting is primarily based on the practice of airing radio advertisements and television advertisements for profit. This is in contrast to public broadcasting, which receives government subsidies and tries to avoid paid advertising interrupting the show. During pledge drives they will interrupt shows to ask for donations.

In the United States, non-commercial educational (NCE) television and radio exists in the form of community radio; however, premium cable services such as HBO and Showtime generally operate solely on subscriber fees and do not sell advertising. This is also the case for the portions of the two major satellite radio systems that are produced in-house (mainly music programming).

Radio broadcasting originally began without paid commercials. As time went on, however, advertisements seemed less objectionable to both the public and government regulators and became more common. While commercial broadcasting was unexpected in radio, in television it was planned due to commercial radio's success. Television began with commercial sponsorship and later transformed to paid commercial time. When problems arose over patents and corporate marketing strategies, regulatory decisions were made by the Federal Communications Commission (FCC) to control commercial broadcasting.

Paid Programming

Commercial broadcasting overlaps with paid services such as cable television, radio and satellite television. Such services are generally partially or wholly paid for by local subscribers and is known as leased access. Other programming (particularly on cable television) is produced by companies operating in much the same manner as advertising-funded commercial broadcasters, and they (and often the local cable provider) sell commercial time in a similar manner.

The FCC's interest in program control began with the chain-broadcasting investigation of the late 1930s, culminating in the "Blue Book" of 1946, *Public Service Responsibility For Broadcast Licensees*. The Blue Book differentiated between mass-appeal sponsored programs and unsponsored "sustaining" programs offered by the radio networks. This sustained programming, according to the Blue Book, had five features serving the public interest:

- Sustaining programs balanced the broadcast schedule, supplementing the soap operas and popular-music programs receiving the highest ratings and most commercial sponsors
- They allowed for the broadcast of programs which, by their controversial or sensitive nature, were unsuitable for sponsorship
- They supplied cultural programming for smaller audiences
- They provided limited broadcast access for non-profit and civic organizations
- They made possible artistic and dramatic experimentation, shielded from the pressures of short-run rating and commercial considerations of a sponsor.

Commercial time has increased 31 seconds per hour for all prime time television shows. For example, ABC has increased from 9 minutes and 26 seconds to 11 minutes and 26 seconds.

Ratings

Programming on commercial stations is more ratings-driven—particularly during periods such as sweeps in the US and some Latin American countries.

Other Factors

Commercial broadcasting (especially free-to-air) is sometimes controversial. One reason is a perceived lack of quality and risk in the programming (to which more conservative elements respond that it is too risqué much of the time), an excessively high ratio of advertising to program time (especially on children's television), and a perceived failure to serve the local interest due to media consolidation. Commercial radio (in particular) is criticized for a perceived homogeneity in programming, covert politically motivated censorship of content, and a desire to cut costs at the expense of a station's identifiable personality. Politics is a major force in media criticism, with an ongoing debate (especially in the United States) as to what moral standards – if any – are to be applied to the airwaves.

Global Commercial Broadcasting

Americas

Commercial broadcasting is the dominant type of broadcasting in the United States and most of Latin America. "The US commercial system resulted from a carefully crafted cooperation endeavor by national corporations and federal regulators."

The best-known commercial broadcasters in the United States today are the ABC, CBS, Fox and NBC television networks and the iHeartMedia radio network, based in the United States. Major cable television in the United States operators include Comcast, Cox Communications and Time Warner Cable. Direct-broadcast satellite (DBS) services include DirecTV and Dish Network.

In an hour of broadcast time on a commercial broadcasting station, 10 to 20 minutes are typically devoted to advertising. Advertisers pay a certain amount of money to air their commercials, usually based upon program ratings or the audience measurement of a station or network. This makes commercial broadcasters more accountable to advertisers than public broadcasting, a disadvantage of commercial radio and television.

Europe

In Europe, commercial broadcasting coexists with public broadcasting (where programming is largely funded by broadcast receiver licences, public donations or government grants).

In the UK, British Sky Broadcasting (BSkyB) is available and WorldSpace Satellite Radio was available.

Asia

One of the best-known commercial services in Asia is the oldest radio station in the region, Radio Ceylon.

Public Broadcasting

Public broadcasting includes radio, television and other electronic media outlets whose primary mission is public service. In much of the world, funding comes from the government, especially via annual fees charged on receivers. In the United States, public broadcasters may receive some funding from both federal and state sources, but generally most financial support comes from underwriting by foundations and businesses ranging from small shops to corporations, along with listener contributions via pledge drives. The great majority are operated as private not-for-profit corporations.

Public broadcasting may be nationally or locally operated, depending on the country and the station. In some countries, public broadcasting is run by a single organization. Other countries have multiple public broadcasting organizations operating regionally or in different languages. Historically, public broadcasting was once the dominant or only form of broadcasting in many countries (with the notable exception of the United States). Commercial broadcasting now also exists in most of these countries; the number of countries with only public broadcasting declined substantially during the latter part of the 20th century.

Defining Public Broadcasting

The primary mission of public broadcasting that of public service, speaking to and engaging as a citizen. The British model has been widely accepted as a universal definition. The model embodies the following principles:

- Universal geographic accessibility
- Universal appeal
- Attention to minorities
- Contribution to national identity and sense of community
- Distance from vested interests
- Direct funding and universality of payment
- Competition in good programming rather than numbers
- Guidelines that liberate rather than restrict

While application of certain principles may be straightforward, as in the case of accessibility, some of the principles may be poorly defined or difficult to implement. In the context of a shifting national identity, the role of public broadcasting may be unclear. Likewise, the subjective nature of good programming may raise the question of individual or public taste.

Within public broadcasting there are two different views regarding commercial activity. One is that public broadcasting is incompatible with commercial objectives. The other is that public broadcasting can and should compete in the marketplace with commercial broadcasters. This dichotomy is highlighted by the public service aspects of traditional commercial broadcasters.

Public broadcasters in each jurisdiction may or may not be synonymous with government controlled broadcasters. In some countries like the UK public broadcasters are not sanctioned by government departments and have independent means of funding, and thus enjoy editorial independence.

Economics

Public broadcasters may receive their funding from an obligatory television licence fee, individual contributions, government funding or commercial sources. Public broadcasters do not rely on advertising to the same degree as commercial broadcasters, or at all; this allows public broadcasters to transmit programmes that are not commercially viable to the mass market, such as public affairs shows, radio and television documentaries, and educational programmes.

One of the principles of public broadcasting is to provide coverage of interests for which there are missing or small markets. Public broadcasting attempts to supply topics of social benefit that are otherwise not provided by commercial broadcasters. Typically, such underprovision is argued to exist when the benefits to viewers are relatively high in comparison to the benefits to advertisers from contacting viewers. This frequently is the case in undeveloped countries that normally have low benefits to advertising.

Cultural Policy

Additionally, public broadcasting may facilitate the implementation of a cultural policy (an industrial policy and investment policy for culture). Examples include:

- The Canadian government is committed to official bilingualism (English and French). As a result, the public broadcaster, the CBC employs translators and journalists who speak both official languages and it encourages production of cross-cultural material. Quebec separatists argue that this is also a policy of cultural imperialism and assimilation.
- In the UK, the BBC supports multiculturalism and diversity, in part by using on-screen commentators and hosts of different ethnic origins. There are also Welsh, Irish and Scottish Gaelic language programmes for the home nations, an Asian Network broadcasting in English and five major languages of South Asia, and the BBC World Service broadcasts in 31 international languages, also funded independently of government.
- In New Zealand, the public broadcasting system provides support to Māori broadcasting, with the stated intention of improving their opportunities, maintaining their cultural heritage and promoting their language.
- In Australia, the Australian Broadcasting Corporation is legally required to 'encourage and promote the musical, dramatic and other performing arts in Australia' and 'broadcasting programs that contribute to a sense of national identity' with specific emphasis on regional and rural Australia'. Furthermore, the Special Broadcasting Service (SBS) is intended to reflect the spirit and sense of multicultural richness and the unique international cultural values within Australian society.

Vested Interests

Public broadcasting, unless its independence is vigorously upheld, can become a tool of government. Similarly, private networks can promote the policies of their owners and suppress other viewpoints, alleging it is in the public interest. Conflicts are common; the state alleges that private broadcasters are attacking it in the interests of the owners, and controls them or takes them over.

Implementation of Public Broadcasting Around the World

The model, established in the 1920s, of the British Broadcasting Corporation—an organization widely trusted, even by citizens of the Axis Powers during World War II—was widely emulated throughout Europe, the British Empire, and later the Commonwealth. The public broadcasters in a number of countries are basically an application of the model used in Britain.

Modern public broadcasting is often a mixed commercial model. For example, the CBC has always relied on a subsidy from general revenues of the government, in addition to advertising revenue, to support its television service. This means they must compete with commercial broadcasting. Some argue that this dilutes their mandate as truly public broadcasters, who have no commercial bias to distort their presentation.

Asia

Hong Kong

In Hong Kong, the Radio Television Hong Kong (RTHK) is the sole public service broadcaster.

Although a government department under administrative hierarchy, it enjoys editorial independence. It operates seven radio channels, and produces television programmes and broadcast on commercial television channels, as these channels are required by law to provide time slot for RTHK television programmes. RTHK would be assigned a digital terrestrial television channel within 2013 to 2015, when the new broadcasting building is completed in Tseung Kwan O.

India

In India, Prasar Bharati is India's public broadcaster. However it does have commercial advertising on it. It is an autonomous corporation of the Ministry of Information and Broadcasting (India), Government of India and comprises the Doordarshan television network and All India Radio. Prasar Bharati was established on 23 November 1997, following a demand that the government owned broadcasters in India should be given autonomy like those in many other countries. The Parliament of India passed an Act to grant this autonomy in 1990, but it was not enacted until 15 September 1997.

Japan

In Japan, the main public broadcaster is the NHK (Japan Broadcasting Corporation). The broadcaster was set up in 1926 and was modelled on the British Broadcasting Company, the precursor to the British Broadcasting Corporation created in 1927. Much like the BBC, NHK is funded by a "receiving fee" by every Japanese household, with no commercial advertising and the maintenance of a position of strict political impartiality. However, rampant non-payment by a large amount of households has led the receiving fee to become something of a political issue. NHK runs two national terrestrial TV stations (NHK General and NHK Educational) and two satellite only services (NHK BS1 and NHK BS Premium services). NHK also runs 3 national radio services and a number of international radio and television services, akin to the BBC World Service. NHK has also been an innovator in television, developing the world's first high definition television technology in 1964 and launching high definition services in Japan in 1981.

Macau

In Macau, Teledifusão de Macau TDM is the public service broadcasting company. The firm was established in January 1982 and was modelled on Rádio e Televisão de Portugal RTP. TDM has two independent editorial arms: the Chinese news channel and the Portuguese news channel, each of which may have different information sources, points of view, and priorities of news. TDM runs both television and radio services. In addition, the Chinese national anthem is not played on TDM Portuguese channel, despite the fact that the entire company is a subject of the Macau SAR government.

Malaysia

In Malaysia, the public broadcaster is the state-owned Radio Televisyen Malaysia (RTM). RTM was previously funded publicly through money obtained from television licensing, however it is currently state subsidised as television licences have been abolished.

At present, RTM operates 8 national, 16 state and 7 district radio stations as well as 2 national terrestrial television channels called TV1 and TV2. RTM has also done test transmissions on a new digital television channel called RTMi. Tests involving 2000 residential homes in the Klang Valley began in September 2006 and ended in March 2007.

Nepal

History of public broadcasting in Nepal started from 1951.

Pakistan

In Pakistan, the public broadcasters are the state-owned Pakistan Broadcasting Corporation (PBC), also known as Radio Pakistan and Pakistan Television. In the past Radio Pakistan was partly funded through money obtained from License fees. In 1999, the Nawaz Sharif government abolished license fees for Radio Pakistan and also abolished its tax exempt status protected under PBC Act 1973. The license fees for Pakistan Television continued. The license fees collection for PTV was given to WAPDA during Musharraf government. Currently WAPDA collects Rs. 35 per house hold on electricity bills as PTV license fees. Television Broadcasting started in Pakistan with a small pilot TV Station established at Lahore Radio from where transmission was first beamed in Black & White with effect from 26 November 1964. Television centres were established in Dhaka, Karachi and Rawalpindi/Islamabad in 1967 and in Peshawar and Quetta in 1974. PTV has various channels transmitting throughout the world including PTV National, PTV World, PTV 2, PTV Global, PTV Bolan etc. Radio Pakistan has stations covering all the major cities, it covers 80% of the country serving 95.5 Million listeners. It has world service in eleven languages daily.

Philippines

The Philippines' primary state television broadcaster is People's Television Network (PTV). Created in 1974 as Government Television (GTV), PTV is no longer state subsidised except for a one-time equity funding for capital outlay in 1992. Aside from PTV, the other public broadcaster is the Intercontinental Broadcasting Corporation (IBC), which the government has already put up for sale. The government no longer holds a controlling interest in the former state broadcaster, Radio Philippines Network (RPN).

The Philippine Broadcasting Service (PBS) is the country's sole state radio broadcaster. Established in 1933 as KZFM by the American colonial Insular Government, the radio station was passed to the Philippine government after the country became independent in 1946.

South Korea

South Korea's representative public broadcasting television network is said to be Korean Broadcasting System, or KBS. Originally a government-controlled channel, it is now an independent broadcasting system. KBS began broadcasting radio in 1947 and opened up to television industry in 1961, and officially founded its own identity by 1973. Another public broadcasting channel is the Munhwa Broadcasting Corporation, or MBC. Known to be the second channel of the country, it also shares the roles of being national television with the KBS.

South Korea also has another public broadcasting channel called Educational Broadcasting System, or EBS. Originally considered to be an extension channel of the KBS, it was spun-off as an educational channel but retained its public broadcasting tasks. KBS and EBS mainly are funded by the commercials that they provide in their channels, but due to growing competitions, this is becoming a problem for them as well.

Europe

Inter-European cooperation of public TV

In most countries in Europe, state broadcasters are funded through a mix of advertising and public finance, either through a licence fee or directly from the government.

Albania

Radio Televizioni Shqiptar (RTSH) is the public broadcaster in Albania.

Analogue TV

- Televizioni Shqiptar (TVSH) is the name of the first public channel of Albania. The domestic TV program is distributed analogically throughout the country and digitally in Tirana through RTSH HD.
- TVSH 2 is the second public TV channel dedicated mainly to sports and live events launched in 2003.

Digital TV

- RTSH HD a digital channel launched in 2012 broadcasts TVSH shows in high-definition quality.
- RTSH Sport
- RTSH Muzikë
- RTSH Art

- TVSH Sat, is the international version of the domestic program broadcast to Europe free to air via satellite.

Radio

- Radio Tirana (also, Radio Tirana 1) is the name of Albania's first radio program, concentrating on news, talk, and features
- Radio Tirana 2 is the name of the second radio program, broadcasting chiefly music and targeted at youth
- Radio Tirana 3 (Programi i Tretë, Radio Tirana International) is the name of the third program, broadcasting the international service on shortwave in Albanian, English, French. Greek, German, Italian, Serbian, and Turkish

Regional

- Radio Televizioni Gjirokastra is the local version of RTSH in Gjirokastër
- Radio Televizioni Korça is the local version of RTSH in Korçë
- Radio Kukësi is the local version of RT in Kukës
- Radio Shkodra is the local version of RT in Shkodër

Belgium

Belgium has three networks, one for each linguistic community:

- RTBF, French-language, with 4 TV channels (La Une, La Deux, La Trois and Arte Belgique) and 6 radios.
- VRT, Dutch-language, with 3 TV channels and 10 radios.
- BRF, German-language, with 1 TV channel and 2 radios.

Originally named INR—*Institut national belge de radiodiffusion* (Dutch: *NIR—Belgisch Nationaal Instituut voor de Radio-omroep*)—the state-owned broadcasting organization was established by law on 18 June 1930. Television broadcasting from Brussels began in 1953, with two hours of programming each day. In 1960 the INR was subsumed into RTB (French: *Radio-Télévision Belge*) and BRT (Dutch: *Belgische Radio- en Televisieomroep*).

On 1 October 1945 INR-NIR began to broadcast some programs in German. In 1961 RTB-BRT began a German-language radio channel, broadcasting from Liège.

In 1977, following Belgian federalization and the establishment of separate language communities, the French-language section of RTB-BRT became RTBF (French: *Radio-Télévision Belge de la Communauté française*), German-language section became BRF (German: *Belgischer Rundfunk*) and Dutch-language stays BRT.

BRT was renamed in 1991 to BRTN (Dutch: *Belgische Radio- en Televisieomroep Nederlandstalige Uitzendingen*) and again in 1998 to VRT (Dutch: *Vlaamse Radio- en Televisieomroeporganisatie*).

Croatia

Croatian Radiotelevision (Croatian: *Hrvatska radiotelevizija, HRT*) is a Croatian public broadcasting company. It operates several radio and television channels, over a domestic transmitter network as well as satellite. As of 2002, 70% of HRT's funding comes from broadcast user fees with each house in Croatia required to pay 79 HRK, kuna, per month for a single television (radio device, computer or smartphone), with the remainder being made up from advertising.

Czech Republic

Czech Television and Czech Radio are national public broadcasting companies in the Czech Republic. Czech Television broadcasts from three studios in Prague, Brno and Ostrava and operates six TV channels ČT1, ČT2, ČT24, ČTSport, ČT :D and ČT Art. Czech television is funded through monthly fee 135 CZK which every household has to pay. Since October 2011 advertising on Czech TV was restricted from four to two channels, namely ČT2 and ČTSport. Czech Radio broadcasts four nationwide stations Radiožurnál, Dvojka, Vltava and Plus, several regional and specialised digital stations. Czech Radio 7 – Radio Prague broadcasts abroad in six languages. Czech Radio is funded throung monthly fee 45 CZK. In the Czech Republic, there is also Czech News Agency (ČTK), a public corporation established by law. The state is not responsible for any ČTK obligations and ČTK is not responsible for any state obligations.

Denmark

DR is the national public service broadcaster. The organization was founded in 1925, on principles similar to those of the BBC in the United Kingdom. Danmarks Radio runs six nationwide television channels and a long lists of radio channels. Financing comes from yearly license fees, that everyone who owns either a television set, a radio, a computer with internet access or a mobile phone, that can access the internet, has to pay.

Faroe Islands

Kringvarp Føroya is the organization in Faroe Islands with public service obligations. Formed in 1957 as a radio broadcaster Útvarp Føroya. Merged with Sjónvarp Føroya TV station on 1 January 2007 to form Kringvarp Føroya. Funded by license fees.

Estonia

ERR (Estonian Public Broadcasting) organizes the public radio and television stations of Estonia. ETV (Estonian Television), the public television station, made its first broadcast in 1955, and together with its sister channel ETV2 has ca. 20% audience share.

Finland

The Finnish Broadcasting Company, or Yleisradio (in Finnish) and Rundradion (in Swedish), abbreviated to Yle (pronounced /yle/), is Finland's national public-broadcasting company. Founded in 1926, it is a public limited company majority owned by the Finnish state, employing around 3,200 people.

France

After World War II, the RTF (Radiodiffusion-télévision française – French television and radio broadcasting) was created to operate the only two channels of television in France. The RTF was transformed into ORTF (Office de radiodiffusion-télévision française – French television and radio diffusion office), a more independent structure, in 1964. The ORTF saw the birth of a third channel in 1972, two years before the dissolution of the structure in 1974. From this date to 2000, each channel had its own direction structure. The first channel (TF1) was sold to the private sector in 1987 (in these years, the channel with the most audience was the other public channel Antenne 2). In 1986 a French/German public channel was created, ARTE, originally broadcast on cable and satellite, the fall of the private channel La Cinq freed some frequencies that it had used each day after 19:00. In 1994 a new public channel, La cinquième was created to use the remaining time on the same frequencies. La cinquième and ARTE subsequently shared the same channels with the exception of satellite, cable, and internet channels where both could be broadcast all day long. In 2000 all the public channels were united into one structure, France Télévisions.

Germany

After World War II, when regional broadcasters had been merged into one national network by the Nazis to create a powerful means of propaganda, the Allies insisted on a de-centralised, independent structure for German public broadcasting and created regional public broadcasting agencies that, by and large, still exist today.

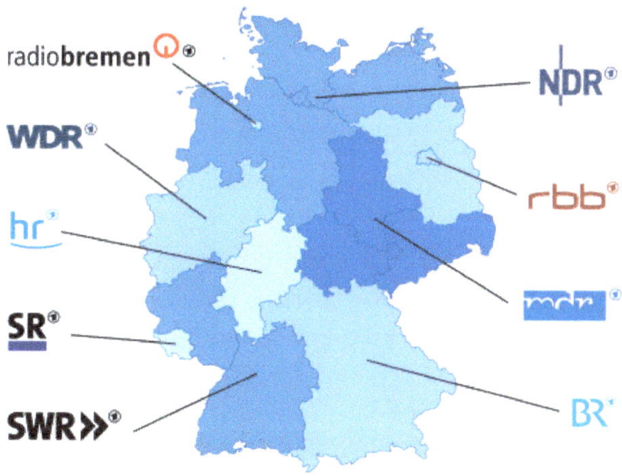

Map of ARD-members

- NDR (Northern Germany), split from former NWDR
- WDR (North Rhine-Westphalia), split from former NWDR
- RBB (Berlin and vicinity), merged from SFB and ORB
- SWR (Germany's south west), merged from SDR and SWF
- BR (Bavaria)
- MDR (Germany's south east), established in 1991

- and the smaller hr (Hesse), SR (Saar) and RB (Land Bremen)

In addition to these nine regional radio and TV broadcasters, which cooperate within ARD, a second national television service—actually called Second German Television (German: *Zweites Deutsches Fernsehen*, ZDF)—was later created in 1961 and a national radio service with two networks (Deutschlandradio) emerged from the remains of Cold War propaganda stations in 1994. All services are mainly financed through license fees paid by every household and are governed by councils of representatives of the "societally relevant groups". Public TV and radio stations spend about 60% of the ~10bn € spent altogether for broadcasting in Germany per year, making it the most expensive public broadcasting system in the world.

The Hans-Bredow-Institut, or Hans-Bredow-Institute for Media Research at the University of Hamburg (HBI) is an independent non-profit foundation with the mission on media research on public communication, particularly for radio and television broadcasting (including public service media providers) and other electronic media, in an interdisciplinary fashion.

In Germany also foreign public broadcasters exist. These are AFN for US-military staff in Germany, BFBS for British military staff, Voice of Russia, RFE and Radio Liberty.

Greece

Hellenic Broadcasting Corporation is the state-owned public broadcaster in Greece. It broadcasts four television channels, ERT1, ERT2, ERT3, ERT HD, as well as five national (ERA 1, ERA 2, ERA 3, Kosmos, ERA Sport), and 18 local radio stations. All national television and radio stations are broadcast through ERT digital multiplexes across the country.

Iceland

Ríkisútvarpið (RÚV) ("The Icelandic National Broadcasting Service") is Iceland's national public-service broadcasting organization. RÚV began radio broadcasting in 1930 and its first television transmissions were made in 1966. In both cases coverage quickly reached nearly every household in Iceland. RÚV is funded by a television licence fee collected from every income taxpayer, as well as advertising revenue. RÚV has been a full active member of the European Broadcasting Union since 1956.

RÚV—which by the terms of its charter is obliged to "promote the Icelandic language, Icelandic history, and Iceland's cultural heritage" and "honour basic democratic rules, human rights, and the freedom of speech and opinion"—carries a substantial amount of arts, media, and current affairs programming, in addition to which it also supplies general entertainment in the form of feature films and such internationally popular television drama series as *Lost* and *Desperate Housewives*. RÚV's lineup also includes sports coverage, documentaries, domestically produced entertainment shows, and children's programming.

Ireland

In Ireland there are two state owned public service broadcasters, RTÉ and TG4. RTÉ was established in 1960 with the merger of *Raidió Éireann* (1926) and *Teilifís Éireann* (1960). TG4 was formed as a subsidiary of *RTÉ* in 1996 as *Teilifís na Gealige (TnaG)*, it was re-branded as *TG4* in 1999, and was made independent of *RTÉ* in 2007.

Both Irish public service broadcasters receive part of the licence fee, with RTÉ taking the lion's share of the funding. Advertising makes up 50% of RTÉ's income and just 6% of TG4's income. 7% of the licence fee is provided to the *Broadcasting Authority of Ireland* since 2006. Up to 2006 the licence fee was given entirely to *RTÉ*.

RTÉ offers a range of free to air services on television; RTÉ 1, RTÉ 2, RTÉjr, and RTÉ News Now. On radio; RTÉ Radio 1, RTÉ 2FM, RTÉ Lyric FM, and RTÉ Raidió na Gaeltachta, as well as a number of channels on DAB.

The Sound and Vision Fund is operated by the Broadcasting Authority of Ireland, this fund receives 7% of the licence fee. The fund is used to assist broadcasters to commission public service broadcast programming. It is open to all independent producers provided they get a free to air or community broadcaster's backing, including TV3, Today FM, BBC Northern Ireland, RTÉ, Channel 4, UTV etc. Pay TV broadcaster Setanta Sports have also received funding for programming through the sound and vision fund provided they provide those shows on a free-to-view basis.

TG4 is an independent Irish language public service broadcaster that is funded by the government through subsidy, the licence fee and through advertising revenue.

TV3 is the only independent broadcaster that has public service commitments.

Italy

RAI brand

The Italian national broadcasting company is RAI – Radiotelevisione Italiana, born as URI in 1924. RAI transmits on digital television on fifteen channels: Rai 1, Rai 2, Rai 3, Rai 4, Rai 5, Rai News 24, Rai Movie, Rai Premium, Rai Sport 1, Rai Sport 2, Rai Storia, Rai Gulp, Rai YoYo, Rai Scuola and Rai HD. RAI also broadcasts via satellite and is involved in radio, publishing and cinema. RAI has the largest audience share (45%) of any Italian television network. Proceeds derive from a periodical standing charge (113.5 euros for each household in 2014) and from advertising. The main competitors of RAI are Mediaset, the biggest national private broadcaster, divided in three channels, and La7, owned by Cairo Editore.

Malta

Public Broadcasting Services (PBS) is the national broadcaster of Malta. It operates three television services (TVM, TVM2, and Parliament TV) and three radio services (Radju Malta, Radju Malta 2, and Magic Malta).

Moldova

Teleradio-Moldova (TRM) is the public funded radio-TV broadcaster in Moldova. It owns the TV channels Moldova 1 and TVMI, and the radio channels Radio Moldova and Radio Moldova Internațional.

Montenegro

RTCG (Radio Television of Montenegro) is the public broadcaster in Montenegro.

Netherlands

The Netherlands uses a rather unusual system of public broadcasting. Public-broadcasting associations are allocated money and time to broadcast their programmes on the publicly owned television and radio channels. The time and money is allocated in proportion to their membership numbers. The system is intended to reflect the diversity of all the groups composing the nation.

Poland

The public broadcasters are Telewizja Polska (TVP) television and Polskie Radio. TVP operates three main channels: TVP 1, TVP 2 and TVP Regionalna. It also broadcasts several digital channels (including TVP 1, TVP1 HD, TVP 2, TVP2 HD, TVP Info, TVP Kultura, TVP Historia, TVP HD, TVP Polonia, TVP Sport, TVP Seriale) via satellite and digital terrestrial television system, and 16 regional affiliates (known as TVP Regionalna, regional channels cooperate when creating most of materials). TVP also runs also services with news, a video streaming (video on demand) service as well as live streaming of all its channels. Polskie Radio operates four nationwide radio channels (which are also available via the broadcaster's website). There are also 17 state-owned radio stations broadcasting in particular regions. TVP and Polish Radio are funded from several sources: state funding, advertising, obligatory tax on all TV and radio receivers, and money from authors/ copyright associations. The public broadcasters offer a mix of commercial shows and programmes they are, by law, required to broadcast (i.e., non-commercial, niche programmes; programmes for children; programmes promoting different points of view and diversity; programmes for different religious and national groups; live coverage of the parliament's session on its dedicated channel: TVP Parlament; etc.). It has to be politically neutral, although in the past there have been cases of political pressure on TVP and Polskie Radio from the governing party. Recently, a new law has been passed by the ruling Law & Justice party, that in public perception allowed the party to take a much larger control over the media that has been possible before. The party states this law to be the first step to a complete public media overdo. Many worry no such improvements are actually coming and that these recent laws are only another step in taking control over the whole country by the Law & Justice party.

There is an ongoing debate in Poland about the semi-commercial nature of TVP and PR. Many people fear that making them into totally non-commercial broadcasters would result in the licence fee payable by households being increased, and fewer people being interested in programmes they offer; others say that TVP in particular is too profit-driven and should concentrate on programming that benefits the society. Some say that state contributions should be programme-based, rather than channel/broadcaster-based, so as all the broadcasters can get them (in this scenario TVP would be privatised and equal to the private TV channels).

Portugal

RTP brand

In Portugal, the national public broadcaster is Rádio e Televisão de Portugal (RTP), which started in 1957 the regular emissions of its first channel, now RTP1. In 1968 appeared the second channel, then called "*segundo programa*", now RTP2. In the 1970s, TV arrived to the Portuguese islands of Madeira and the Azores, with the creation of two regional channels: RTP Madeira in 1972 and RTP Açores in 1976.

Until the 1990s, the state owned the monopoly of TV broadcasting, so RTP1 and RTP2 were the only Portuguese channels, both with similar generalist programmes. In 1990, RTP1 was renamed "Canal 1", and in 1992 RTP2 was renamed "TV2". With the creation of the two private channels, SIC in 1992 and Televisão Independente in 1993, the philosophy of the public service changed: in 1995, *TV2* was again renamed RTP2 and became an alternative channel dedicated to culture, science, arts, documentaries, sports (without football), minorities and children. Since then, RTP2 has no advertising. Canal 1, renamed back RTP1 also in 1995, remained the commercial channel of RTP group, focused on entertainment, information and major sport competitions. In 2004, after a great reestructuration period, RTP started its currint branding. That year was also created the two thematic channels of the group—RTPN, 24 hour-news channel which name changed to RTP Informação in 2011; and RTP Memória, dedicated to classic RTP programming. In 2014 the headquarters of RTP2 were transferred from Lisbon to Porto. The group also has two international channels: RTP Internacional, founded in 1992 and dedicated to Europe, Asia and Americas, and RTP África, founded in 1998 focused in Africa, mainly in the CPLP countries of Angola, Cape Verde, Guinea-Bissau, Mozambique and São Tomé and Príncipe.

The RTP group is financed by the advertising revenues from RTP1, RTP Informação, RTP Memória, RTP África, and RTP Internacional, and also by the *taxa de contribuição audiovisual* (broadcasting contribution tax), which is incorporated in electricity bills. Funding from the government budget ceased in 2014, during the Portuguese financial crisis.

Romania

Romanian Television (TVR) is the national public TV broadcaster in Romania. It operates five channels: TVR1, TVR2, TVR3, TVRi, and TVR HD, along with six regional studios in Bucharest, Cluj-Napoca, Ia□i, Timi□oara, Craiova, and Târgu Mure□.

The public radio broadcaster is Romanian Radio Broadcasting Company (Radio Romania). It operates FM and AM, and internet national, regional, and local radio channels. The regional and local stations are branded as Radio România Regional. Broadcasting in 12 languages, Radio Romania International is the company's international radio station.

TVR and Radio Romania are funded through a hybrid financing system, drawing from the state budget, a special tax (incorporated in electricity bills), and advertising too.

Nordic Countries

National public broadcasters in The Nordic countries were modeled after the BBC and established a decade later: Radioordningen (now DR) in Denmark, Kringkastingselskapet (now NRK) in Norway, and Radiotjänst (now Sveriges Radio and Sveriges Television) in Sweden (all in 1925). In 1926 Yleisradio (Swedish: Rundradion) now YLE was founded in Finland. All four are funded from television licence fees costing (in 2007) around €230 (US$300) per household per year.

Serbia

Radio Television of Serbia (RTS) is the national public broadcaster in Serbia. It operates a total of five television channels (RTS1, RTS2, RTS Digital, RTS HD and RTS SAT) and five radio stations (Radio Belgrade 1, Radio Belgrade 2, Radio Belgrade 3, Radio Belgrade 202, and Stereorama). RTS is primarily funded through public television licence fees bundled with electricity bills paid monthly, as well as advertising.

Slovakia

RTVS logo

Radio and Television Slovakia (RTVS) is the national public broadcaster in Slovakia, with headquarters in Bratislava. This organisation was created in 2011 by merger of Slovak Television and Slovak Radio. RTVS broadcasts two television channels (STV1, STV2), five FM radio stations (Rádio Slovensko, Rádio Devín, Rádio Regina, Rádio_FM and Rádio Patria), one satellite radio channel (Radio Slovakia International) and three digital only radio stations (Rádio Klasika, Rádio Litera and Rádio Junior). RTVS is funded through monthly fee of €4.64, which every household with electricity connection must pay. The director of RTVS is Václav Mika. RTVS is a full member of European broadcasting union.

Spain

In Spain, being a highly decentralised country, two public broadcasting systems coexist: a national broadcasting television, Radio y Televisión Española (RTVE), that can be watched all around Spain, and many autonomic TV channels, only broadcast within their respective Autonomous Community. Televisión Española was founded in 1956, during Franco's dictatorship. It broadcasts two different TV-channels: TVE1 (a.k.a. *La Primera* or *La uno*), that is a wide-range audience general channel; and TVE2, (a.k.a. *La dos*), that tends to offer cultural programation, as well as sport competitions. Till 2008, RTVE was funded both from public sources and private advertising; however, the Spanish government has recently decreed that, starting in September 2009, RTVE's channels shall be funded with taxpayer's money and with private founding raised from the rest of Spain's private TV stations, thus removing advertising from the broadcaster. A TV licence fee has been suggested, but with little popular success.

Moreover, each of the autonomous communities of Spain have their own public broadcaster, all of them members of FORTA, usually consisting in either one or two public channels that tend to reproduce the model set up by Televisión Española: a general channel and a more cultural related one. In the Autonomous Communities that have their own official language besides Spanish, those channels may broadcast not in Spanish, but in the other co-official language. For example, this occurs in Catalonia, where Televisió de Catalunya broadcasts mainly in Catalan. In the Basque Country, Euskal Telebista (ETB) has three channels, two of which broadcast only in basque (ETB 1 and ETB 3), whereas the other (ETB 2) broadcasts in Spanish. In Galicia, the Television de Galicia and the G2. All the autonomic networks are publicly founded, and also admit private advertising.

Sweden

The logo of SVT1

The logo of Sveriges Radio

Sweden has three Public Service broadcasters and those are SVT (Sveriges Television), UR (Utbildningsradion) and SR (Sveriges Radio). SVT is the national public television Broadcaster with 4 channels (SVT 1, SVT 2, SVT Barnkanalen and SVT 24). The aim is to make programs for everybody. For example, Sweden has their original population called the Sami People and SVT make programs in their language for them to watch. There are also Finnish people in Sweden, so SVT sometimed show the news in Suomi, the program is called SVT Uutiset. SR has the same purpose as SVT but SR is the Swedish national radio broadcasting station with the channels, P1, P2, P3, P4, P5 and the Finnish channel SR sisuradio.

United Kingdom

The United Kingdom has a strong tradition of public service broadcasting. In addition to the BBC, established in 1922, there is also Channel 4, a publicly owned, commercially funded public service broadcaster, and S4C, a Welsh-language broadcaster in Wales. Furthermore, the two commercial broadcasters ITV and Channel 5 also have significant public service obligations imposed as part of their licence to broadcast.

In the UK there are also small community broadcasters. There are now 228 stations with FM broadcast licences (licensed by Ofcom). Community radio stations typically cover a small geographical area with a coverage radius of up to 5 km and run on a nonprofit basis. They can cater for whole communities or for different areas of interest—such as a particular ethnic group, age group or interest group. Community radio stations reflect a diverse mix of cultures and interests. For example, you can listen to stations catering to urban or experimental music, while others are aimed at younger people, religious communities or the armed forces and their families.

North America

Canada

In Canada, the main public broadcaster is the national Canadian Broadcasting Corporation (CBC; French: *Société Radio-Canada*) which operates two national television networks (CBC Television and Ici Radio-Canada Télé), four radio networks (CBC Radio One, CBC Radio 2, Ici Radio-Canada Première and Ici Musique) and several cable television channels including two 24-hour news channels (CBC News Network and Ici RDI) in both of Canada's official languages and the French-language science channel Ici Explora. CBC's national television operations and some radio operations are funded partly by advertisements, in addition to the subsidy provided by the federal government. The cable channels are commercial entities owned and operated by the CBC and do not receive any direct public funds, however, they do benefit from synergies with resources from the other CBC operations. The CBC has been frequently dealt with budget cuts and labour disputes, often resulting in a debate about whether the service has the resources necessary to properly fulfill its mandate.

In recent years, the CBC has also expanded into new media ventures including the online radio service CBC Radio 3, music streaming service CBC Music, and the launch of online news services, such as CBC Hamilton, in some markets which are not directly served by their own CBC television or radio stations.

In addition, several provinces operate public broadcasters; these are not CBC subentities, but distinct networks in their own right. These include the English-language TVOntario (TVO) and the French-language TFO in Ontario, Télé-Québec in Quebec, and Knowledge Network in British Columbia. TVO and Télé-Québec operate through conventional transmitters and cable, while TFO and Knowledge Network are cable-only channels.

Amherst Island public radio

Canada is also home to a number of former public broadcasting entities that have gone private. CTV Two Alberta, which is licensed as an educational television station in Alberta, was once owned by the Alberta government as the public broadcaster Access, until the channel was sold to CHUM Limited in 1995. To fulfill its license conditions as an educational station, it broadcasts educational and children's programming during daytime hours while airing entertainment programming favoured by advertisers and viewers in primetime. Public radio station CKUA in Alberta was also formerly operated by Access, before being sold to the non-profit CKUA Radio Foundation which continues to operate it as a community-funded radio network. CJRT-FM in Toronto also operated as a public government-owned radio station for many years; while no longer funded by the provincial government, it still solicits most of its budget from listener and corporate donations and is permitted to air only a very small amount of commercial advertising. One television station, CFTU in Montreal, operates as an educational station owned by a private not-for-profit consortium of educational institutions in the province of Quebec called CANAL. City Saskatchewan was once the Saskatchewan Communications Network, a cable-only educational and cultural public broadcaster owned by the government of Saskatchewan, until it was sold to Bluepoint Investment Corporation in 2010 and later to Rogers Media in 2012. Live CTV Two Alberta it airs educational and children's programming during the day as a condition of license.

Some local community stations also operate non-commercially with funding from corporate and individual donors. In addition, cable companies are required to produce a local community channel in each licensed market. Such channels have traditionally aired community talk shows, city council meetings and other locally oriented programming, although it is becoming increasingly common for them to adopt the format and branding of a local news channel.

Canada also has a large number of campus radio and community radio stations.

United States

Early public stations were operated by state colleges and universities, and were often run as part of the schools' cooperative extension services. Stations in this era were internally funded, and did not rely on listener contributions to operate; some accepted advertising. Networks such as Iowa, South Dakota, and Wisconsin Public Radio began in this way.

The Gregory Hall on the campus of University of Illinois at Urbana-Champaign hosted an important meeting of the National Association of Educational Broadcasters in the 1940s, that spawned both PBS and NPR

The concept of a "non-commercial, educational" station *per se* does not show up in U.S. law until the 1940s, when the FM band was moved to its present location; the part of the band between 88.1 and 91.9 MHz is reserved for such stations, though they are not limited to those frequencies. For example, WFIU-Bloomington, Ind. has its FM frequency at 103.7 MHz. Houston's KUHT was the nation's first public television station, and signed on the air on 25 May 1953 from the campus of the University of Houston. This phenomenon continued in other big cities in the 1950s; in rural areas, it was not uncommon for colleges to operate commercial stations instead (e.g., the University of Missouri's KOMU-TV, an NBC affiliate in Columbia).

In the United States, other than a few direct services, public broadcasting is almost entirely decentralized and is not government operated, but does receive some government support. Some of the funding comes from community support to hundreds of public radio and public television stations, each of which is an individual entity licensed to one of several different non-profit organizations, municipal or state governments, or universities. Sources of funding also include on-air and online pledge drives and the sale of underwriting "spots" (typically 15–30 seconds) to sponsors. Public radio and television stations often produce their own programs as well as purchase additional programming from national producers and program distributors such as National Public Radio (NPR), Public Broadcasting Service (PBS), Public Radio International (PRI), American Public Television (APT), American Public Media (APM), and Public Radio Exchange (PRX). U.S. federal government support for public radio and television is filtered through a separate organization, the Corporation for Public Broadcasting (CPB).

Public broadcasting is sometimes also referred to as *public media*, in an effort to capture the expansion of public broadcasting content from radio and television into digital technologies, in particular the web and mobile platforms. While some consider public media to be analogous to public broadcasting, others use the term more broadly to include all noncommercial media.

Individual stations and programs rely on highly varied proportions of funding. Program-by-program funding creates the potential for conflict-of-interest situations, which must be weighed program by program under standards such as the guidelines established by PBS. Donations are widely dispersed to stations and producers, giving the system a resilience and broad base of support but diffusing authority and impeding decisive change and priority-setting.

Television

In the United States, the Public Broadcasting Service (PBS) (formerly National Educational Television) and American Public Television (formerly Eastern Educational Television Network) distribute television programs to a nationwide system of independently owned and operated television stations (some having the term "PBS" in their name) supported largely by state and federal governments as well as viewer support, with commercial underwriters donating to specific programs and receiving a short thanks for their contributions. Such underwriting may only issue declarative statements (including slogans) and may not include "calls to action" (i.e., the station cannot give out prices, comparative statements, or anything that would persuade the listener to patronize the sponsor). Most communities also have Public-access television channels on local cable television systems, which are generally paid for by Cable television franchise fees and sometimes supported in part through citizens donations.

The U.S. government produces one channel for domestic citizen consumption: NASA TV, a suite of channels covering the country's space program and a collection of science education programs. Formerly, there was the DoD News Channel, which distributed news regarding the country's military operations. NASA TV is and the DoD News channel was distributed solely by satellite and Internet and thus do not have the extensive reach of other countries' national broadcasters.

U.S. public broadcasting for television has, from the late 1960s onward, dealt with severe criticism from conservative politicians and think-tanks, which allege that its programming has a leftist bias.

Radio

A public radio network, National Public Radio (NPR), was created in 1970, following the passage of the Public Broadcasting Act of 1967, which established the Corporation for Public Broadcasting. This network is colloquially though inaccurately conflated with *public radio* as a whole, when in fact "public radio" includes many organizations. Some independent local public radio stations buy their programming from distributors such as NPR; Public Radio International (PRI); American Public Media (APM); Public Radio Exchange (PRX); and Pacifica Radio, most often distributed through the Public Radio Satellite System (PRSS). Around these distributed programs, stations fill in varying amounts of local and other programming. A number of public stations are completely independent of these programming services, producing all or most of their content themselves. Public radio stations in the United States tend to broadcast a mixture of news and talk programs along with music and arts/cultural programming. Some of the larger operations split off these formats into separate stations or networks. Music stations are probably best known for playing classical music, although other formats are offered, including the time-honored "eclectic" music format that is rather freeform in nature common among college radio stations. Jazz is another traditional, but declining, public radio programming staple. Cultural Native American and Mexican American music and programming are also featured regionally.

The U.S. government operates some limited direct broadcasting services, but all are either highly specialized (and, since the dawn of the millennium, automated) information services (WWV/WWVH time service, NOAA Weather Radio) or targeted at foreign audiences like Voice of America. From 1948 to 2013, foreign-targeted broadcasts, many of which were intended as propaganda, were barred from U.S. audiences because of the Smith–Mundt Act, a restriction that has since been lifted. While NOAA Weather Radio has individual terrestrial repeaters across the United States (albeit on a special band reserved for such broadcasts), WWV, VOA and others operate from single shortwave facilities; none of these services can be heard on the AM or FM bands most common on U.S. radio.

Local stations derive some of the funding for their operations through regular pledge drives seeking individual and corporate donations, and corporate underwriting. Some stations also derive a portion of their funding from federal, state and local governments and government-funded colleges and universities (in addition to receiving free use of the public radio spectrum). The local stations then contract with program distributors and also provide some programming themselves. NPR produces its own programming. PBS, by contrast, does not create its own content. NPR also receives some direct funding from private donors, foundations, and from the Corporation for Public Broadcasting.

Although most "public" radio stations are noncommercial, not all noncommercial stations are public. Community radio stations may have some independently produced programs in common with their public counterparts but, because they do not rely on government subsidy of any sort, cannot be considered public, even in the American sense of the term.

Oceania

Australia

In Australia, the Australian Broadcasting Corporation (ABC) is owned by the Australian Government and is 100% taxpayer funded. The multicultural Special Broadcasting Service (SBS), another public broadcaster, now accepts limited sponsorship and advertising.

In addition, there is a large Australian community radio sector, funded in part by federal grants via the Community Broadcasting Foundation, but largely sustained via subscriptions, donations and business sponsorship. As of June 2005, there were 442 fully licensed community radio stations (including remote Indigenous services) and a number of community television stations (most operating as Channel 31 despite being unrelated across different states). They are organised similarly to PBS and NPR stations in the United States, and take on the role that public access television stations have in the US.

New Zealand

In New Zealand all broadcasters are given a small slice of public broadcasting responsibility, because of the state-funded agency NZ On Air. This is because of NZ On Air's requirement for public-service programmes across all channels and stations, instead of being put into one single network. The former public broadcaster BCNZ (formerly NZBC – New Zealand Broadcasting Corporation) was broken up into separate state-owned corporations, Television New Zealand (TVNZ) and Radio New Zealand (RNZ). While RNZ remains commercial-free, about 90% of funding for TVNZ comes from selling advertising during programmes on their two stations. TVNZ continues to be a public broadcaster; however like CBC Television in Canada it is essentially a fully commercial network in continuous ratings battles with other stations, which continues to be a controversial issue within New Zealand. With the shutdown of TVNZ 7, the only fully non-commercial public-service network in New Zealand is Radio New Zealand.

Aside from television, New Zealand has a rich public radio culture, Radio New Zealand being the main provider, with a varied network (Radio New Zealand National) and a classical musical network (Radio New Zealand Concert). RNZ also provides the Pacific with its Radio New Zealand International. Aside from RNZ almost all of New Zealand's 16 regions has an "access radio" network. All these networks are commercial-free.

South America

Argentina

State presence in television had a strong history, not in the way of European style public service radio or television. The private sector has taken an active role in the development of television in Buenos Aires. In opposition, state broadcasters tend to be federal and technical innovative, such as the Argentinian Televisión Publica, the first 60 years old national TV station.

Brazil

In Brazil, the two main public broadcasters are the EBC (Empresa Brasil de Comunicação) and the Fundação Padre Anchieta. The EBC was created in 2007 to manage the Brazilian federal government's radio and television stations. EBC owns broadcast networks such as TV Brasil (launched in 2007, being the merger of Rio de Janeiro's TV Educativa (1975-2007) and Brasília's TV Nacional (1960-2007)), the Nacional and MEC radios. The Padre Anchieta Foundation was created by the government of the state of São Paulo in 1967 and includes a television station (TV Cultura, launched in 1969 in São Paulo), and a radio station (Rádio Cultura). The Padre Anchieta Foundation is a privately held company which maintains autonomy.

Chile

Unlike most of Latin America, Chilean television was founded in an attempt to bring public television. Thus, the first TV stations were not commercial, but were part of some of the biggest universities in Chile: the University of Chile (owner of the former Channels 9 and 11 until 1993), the Catholic University of Chile on channels 2 and 13 until 2010, and the Pontifical Catholic University of Valparaíso on channels 8 and 4. Channel 8, in Valparaíso, is the first and oldest station on Chile, transmitting since October 5, 1957. Anyway, as the Universities found themselves in a quite bad economic situation, as soon as 1961 they began transmitting advertisements between their programming, being the first of them the channel 9, showing a Motorola TV set. This kind of disguised advertising took the name of "Payola". This situation, added to the fact that TV was only reaching Santiago and Valparaíso, led to the creation of a state network that should serve all the country. This network, created in 1964 and in operation since October 24, 1969, is known as "Televisión Nacional de Chile". After the military government of Augusto Pinochet, the television was deregulated. Thus, two new commercial channels were born: Megavisión (Channel 9, on October 23, 1990) and La Red (Channel 4, on May 12, 1991). The University of Chile's channel 11 also was privatized on October 1, 1993. This network is known today as "Chilevisión".

The closest model to the British BBC is that of Chile's Televisión Nacional, an open channel serving the entire country (including Easter Island and Antarctica bases). Televisión Nacional, popularly known as channel 7 because of its Santiago frequency, is governed by a seven-member board appointed by both the President and the Senate. It is meant to be independent of political pressures, although accusations of bias have been made, especially during election campaigns.

Ecuador

Ecuador TV is the public service channel of Ecuador established in October 2007. The channel was established at the same time as the installation of the Ecuadorian Constituent Assembly so that the sessions could be transmitted live to all the country.

Venezuela

Recently, under the initiative of the Venezuelan government of president Hugo Chávez, and with the sponsorship of the governments of Argentina, Bolivia, Cuba, Ecuador and Nicaragua, the news and documentary network teleSUR was created with the intended to be an instrument toward the "concretizing of the Bolivarian idea" through the integration of America, and as a counterweight to what the governments that funds it consider a "distorted view of Latin American reality by privately run networks that broadcast to the region". There is an ongoing debate on whether teleSUR will be able become a neutral and fair news channel able to counter the huge influence of global media outlets, or whether it will end up as a propaganda tool of the Venezuelan government, which owns a 51 percent share of said channel.

List of Public Broadcasters

Africa

- SABC–South Africa
- Television
    - SABC 1
    - SABC 2
    - SABC 3
- Radio (Each official language has a station)
    - SAfm (English)
    - RSG (Afrikaans)
    - Ukhozi FM (Zulu)
    - Umhlobo Wenene FM (Xhosa)

    - Thobela FM (Pedi)
    - Lesedi FM (Sotho)
    - Motsweding FM (Tswana)
    - Phalaphala FM (Venda)
    - Munghana Lonene FM (Tsonga)
    - Ligwalagwala FM (Swazi)
    - Ikwekwezi FM (Ndebele)
    - X-K FM (!Xu and Khwe)—These are not official languages but significant communities exist in the Northern Cape Province
- Botswana TV—Botswana
- GBC—Ghana
- Televisão Pública de Angola—Angola
- Televisão de Moçambique—Mozambique
- SNRT—Morocco
- Établissement public de télévision—Algeria
    - Programme National
    - Canal Algérie
    - A3
    - Chaîne 4
    - Coran TV
- ERTU—Egypt
- LJBC—Libya
- LBS—Liberia
- NTA—Nigeria
- FRCN—Nigeria
- NBC—Namibia
- TBC—Tanzania

Asia

Asian Legends

- Pakistan Television Corporation—Pakistan

- Educational Broadcasting System—South Korea
- Korean Broadcasting System—South Korea
- Munhwa Broadcasting Corporation—South Korea; semi-public, semi-private
- NHK—Japan
- Prasar Bharati—India
    - All India Radio
    - Doordarshan
- Israel Broadcasting Authority—Israel
- IRIB—Iran
- JRTV—Jordan
- Télé-Liban—Lebanon
- Nepal Television (NTV)—Nepal
    - NTV Metro 2
    - Radio Nepal
- Vietnam Television—Vietnam
- People's Television Network—Philippines
- Radio Televisyen Malaysia—Malaysia
- Syrian Television—Syria
- Televisi Republik Indonesia—Indonesia
- Thai Public Broadcasting Service—Thailand
- Abu Dhabi TV—United Arab Emirates
- Al Jazeera—Qatar

Brunei Darussalam

- Radio Televisyen Brunei(RTB)

China

- China Central Television
- Radio Television Hong Kong （香港電台|RTHK） — Hong Kong
- Teledifusão de Macau （澳門廣播電視股份有限公司 "澳廣視" |TDM） -Macau

Indonesia

- Televisi Republik Indonesia (TVRI)

- Radio Republik Indonesia (RRI)

Malaysia

- Radio Televisyen Malaysia (RTM)

North Korea

- Korean Central Television (KCTV)

Singapore

- Media Corporation of Singapore (MediaCorp)

South Korea

- Korean Broadcasting System (KBS)
- Educational Broadcasting System (EBS)

Taiwan

- Taiwan Broadcasting System （台灣公共廣播電視集團|TBS）
- Radio Taiwan International （中央廣播電台|RTI）
- National Education Radio （國立教育廣播電臺|NER）
- Police Radio Station （警察廣播電台|PRS）
- Taipei Broadcasting Station （台北廣播電台|TBS）
- Kaohsiung Broadcasting Station （高雄廣播電台|KBS）

Europe

- 3sat-Germany/Austria/Switzerland
- ARD—working partnership of German public-service broadcasters
    - Westdeutscher Rundfunk, WDR—Cologne
    - Norddeutscher Rundfunk—Hamburg
    - Mitteldeutscher Rundfunk—Leipzig
    - Bayerischer Rundfunk—Munich
    - Südwestrundfunk—Stuttgart
    - Rundfunk Berlin-Brandenburg—Berlin
    - Hessischer Rundfunk—Frankfurt
    - Saarländischer Rundfunk—Saarbrücken
    - Radio Bremen—Bremen
    - German radio and TV from ARD: each broadcaster has several radio programs, like

the biggest one (WDR) more than five "full" radio stations ("1 Live", WDR 2 (pop and info radio), WDR 3 (classical music) ... WDR 5) plus regional radio. WDR produces one TV-programme for its state and contributes to "Das Erste"–TV of ARD and to more TV-channels.

- Arte— France/Germany
- BBC—United Kingdom
    - BBC Alba
    - BBC One
    - BBC Two
    - BBC Four
    - BBC News
    - BBC Parliament
    - BBC HD
    - CBBC
    - CBeebies
    - BBC Radio 1
    - BBC Radio 2
    - BBC Radio 3
    - BBC Radio 4
    - BBC Radio 5 Live
    - BBC Radio 1Xtra
    - BBC Radio 4 Extra
    - BBC Radio 5 Live Sports Extra
    - BBC Radio 6 Music
    - BBC Asian Network
    - BBC World Service
    - BBC Radio Scotland
    - BBC Radio nan Gàidheal
    - BBC Radio Shetland
    - BBC Radio Orkney
    - BBC Radio Wales

  - BBC Radio Cymru
  - BBC Radio Ulster
  - BBC Radio Foyle
  - BBC Local Radio—40 stations
- BBC Worldwide
  - BBC World News
  - BBC Lifestyle
  - BBC Knowledge
  - BBC HD
  - BBC World Service
  - BBC Entertainment
  - BBC America
  - BBC Canada
  - BBC UKTV
- BNT—Bulgaria
- BRF—German-speaking Community of Belgium
- BVN—Flanders and Netherlands television
- Česká televize—Czech Republic
- Channel 4—United Kingdom
- Danmarks Radio—Denmark
- Duna TV—Hungary
- EITB–Basque Country
- EAT (Greek Public Television)—Greece (formerly)
- ERT—Greece (1938 – 11 June 2013; 11 June 2015–present)
- ETV—Estonia
- LRT–Lithuania
- France Télévisions—France
  - France 2
  - France 3
  - France 4

    - France 5
- GBC—Gibraltar
- KNR—Greenland
- HRT—Croatia
- LRT—Lithuania
- LTV—Latvia
- Magyar Televízió—Hungary
- MRT—Macedonia
- NERIT—Greece (formerly; 4 May 2014 – 11 June 2015)
- Netherlands Public Broadcasting—Netherlands
    - AVRO
    - BNN
    - EO
    - KRO
    - LLiNK
    - MAX
    - NCRV
    - TROS
    - VARA
    - VPRO
- NRK—Norway
- NTKU-Ukraine
- ORF—Austria
- OTR (*Общественное Телевидение России*, Obshchestvennoe Televidenie Rossii)—Public Television of Russia
- Polish Radio-Poland
- PBS—Malta
- Radio France—France
    - France Inter
    - France Info

  - France Culture
  - France Musique
  - France Bleu
  - FIP
  - Le Mouv'
- Radio France Internationale—France
- Ràdio i Televisió d'Andorra—Andorra
- Radio România
- RAI—Italy
- RAS—Italy
- RTBF—Wallonia, Belgium
- RTCG—Montenegro
- RTÉ—Ireland
  - RTÉ 1
  - RTÉ 2
  - RTÉjr
  - RTÉ News Now
  - RTÉ Radio 1
  - RTÉ 2FM
  - RTÉ Lyric FM
  - RTÉ Raidió na Gaeltachta
- RTP—Portugal
  - RTP1
  - RTP2
  - RTP Informação
  - RTP Memória
  - RTP África
  - RTP Internacional
  - RTP Açores
  - RTP Madeira

    - RDP Antena 1
    - RDP Antena 2
    - RDP Antena 3
    - RDP Internacional
    - RDP África
    - RDP Açores
    - RDP Madeira
- RTS—Serbia
- RTSH—Albania
- RTK–Kosovo
- RTV Slovenija—Slovenia
- RTVE—Spain
    - TVE1
    - TVE-2
    - Clan TVE
    - Canal 24 Horas
    - Teledeporte
    - TVE Internacional
- RÚV—Iceland
- S4C—Wales, United Kingdom
- Kringvarp Føroya—Faroe Islands
    - Sjónvarp Føroya
    - Útvarp Føroya
- SRG SSR—Switzerland
    - SRF 1
    - SRF zwei
    - SRF info
    - RTS Un
    - RTS Deux
    - RSI La 1

  - RSI La 2
  - Televisiun Rumantscha
- SMtv—San Marino
- SVT – Sveriges Television—Sweden
- SR – Sveriges Radio—Sweden
- UR—Sweden
- TG4 — Ireland
- TVP—Poland
  - TVP 1
  - TVP 2
  - TVP Info
  - TVP Regionalna
  - TVP Kultura
  - TVP Historia
  - TVP Seriale
  - TVP HD
  - TVP Sport
  - TVP Rozrywka
  - TVP ABC
  - Biełsat TV
- Televisió de Catalunya—Spain
- TVR—Romania
- TRT—Turkey
- VRT—Flanders, Belgium
- YLE – Yleisradio—Finland
- ZDF—Germany

North America

Barbados

- Caribbean Broadcasting Corporation—Saint Michael, Barbados

Costa Rica

- Canal 13 (SINART)—Costa Rica
- Canal 15 (Universidad de Costa Rica)—Costa Rica

Cuba

- Cubavisión—Cuba

Canada

- Canadian Broadcasting Corporation—national
    - CBC Television
    - CBC News Network
    - CBC North
    - Documentary
    - Ici Radio-Canada Télé
    - Réseau de l'information
    - ARTV
    - Explora
    - TV5 Québec Canada
    - CBC Radio One
    - CBC Radio 2
    - CBC Radio 3 (no terrestrial broadcasting)
    - Ici Radio-Canada Première
    - Espace musique
    - Radio Canada International (no terrestrial broadcasting; international broadcasting only)
    - CBC Parliamentary Television Network (former)
    - Bande à part (former)
    - Newsworld International (former; international broadcasting only)
    - Trio (former; international broadcasting only)
    - CBC-2 (proposed)
    - Télé-2 (proposed)
- BBC Worldwide—national

  - BBC Canada
  - BBC Kids
- Knowledge—British Columbia
- TVOntario—Ontario
- Télé-Québec—Quebec
- TFO—Ontario
- CFTU-TV—Montreal, Quebec (non-profit)
- CKUA Radio Network—Alberta (non-profit)
- CJRT-FM—Toronto, Ontario (non-profit)
- Access—Alberta (former)
- Saskatchewan Communications Network—Saskatchewan (former)

United States

- Public Broadcasting Service (October 1970)
- National Public Radio (February 1970)
- New York Public Radio (July 1924)
- Chicago Public Radio (April 1943)
- Pacifica Radio (April 1949)
- Boston Public Radio (October 1951)
- American Public Television (1961)
- American Public Media (January 1967)
- MHz Networks (March 1972)
- Public Radio International (1983)
- World (TV channel) (December 2005)
- Create (TV network) (January 2006)

Oceania

Australia

- Australian Broadcasting Corporation
  - ABC TV
  - ABC2
  - ABC3

  - ABC News 24
  - ABC Local Radio—*54 stations*
  - ABC Radio National
  - ABC NewsRadio
  - ABC Classic FM
  - ABC Triple J
  - ABC Radio Australia
  - ABC Radio Grandstand
- Special Broadcasting Service—*Also available in New Zealand in selected regions*
  - SBS One
  - SBS Two
  - SBS Radio—*Includes SBS Radio One, Two, Pop Asia, and CHILL*

East Timor

- Radio-Televisão Timor Leste

New Zealand

- RNZ
  - RNZ National
  - RNZ Concert
  - RNZ Parliament
- Māori Television
- TVNZ—*six channels, commercial television and public television*
- Triangle TV—*Auckland Region broadcaster*

NZ Local Networks (Can be Included)

- CUETV
- TVHB
- 45 South TV
- Channel North Television
- Access Taranaki 104.4FM
- Access Manawatu 999AM
- Access Wellington 783AM

- Mainland TV Nelson

Oceania International Services

- Australia Network
- RNZ INTERNATIONAL

Pacific Islands

- FBC Fiji
- SBC Samoa
- TBC Tonga
- PBS American Samoa

South America

Argentina

- Canal 7 Argentina—Argentina (nationwide)
- Canal 10 – Tucumán—Tucumán, Argentina
- Canal 10 – Rio Negro—Rio Negro, Argentina
- Canal 10 – Córdoba—Córdoba, Argentina

Brazil

- Empresa Brasil de Comunicação
    - TV Brasil — Rio de Janeiro, Brasília, São Paulo, São Luís and broadcast affiliates
    - TV Brasil Internacional—Available in Africa, Americas, Portugal and Japan by cable and satellite
    - Rádio Nacional—Rio de Janeiro, Brasília and Amazon region
    - Rádio MEC—Rio de Janeiro and Brasília
- Fundação Padre Anchieta
    - TV Cultura—São Paulo and broadcast affiliates.
    - MultiCultura—São Paulo
    - Univesp TV—São Paulo
    - TV Rá-Tim-Bum—Nationwide by cable and satellite.
    - Rádio Cultura—São Paulo

Chile

- Televisión Nacional de Chile—Chile (nationwide)

Colombia

National

- Radio Televisión Nacional de Colombia
    - Señal Colombia
    - Señal Institucional

Regional

- Canal 13
- Teleantioquia
- Telecafé
- Teleislas
- Televisión Regional del Oriente
- Telecaribe
- Telepacífico

Local

- Canal Capital
- Telemedellín

Ecuador

- Ecuador TV—Ecuador http://www.ecuadortv.ec

Guyana

- National Communications Network (NCN)

Uruguay

- TNU – Televisión Nacional de Uruguay
- S.O.D.R.E—Servicio Official Radioletrico (en desuso)
- Radio Clasica
- Radio Uruguay
- Radio Nacional

Venezuela

- Venezolana de Televisión—Venezuela (nationwide)
- Televisión Educativa—Venezuela
- Visión Venezuela—Venezuela

Transnational

- teleSUR—Reaches the entire continent, Europe and Northern Africa. Owned by La Nueva Televisora del Sur, a public company sponsored by several American countries.

Remote Broadcast

In broadcast engineering, a remote broadcast (usually just called a remote or a live remote, or in news parlance, a live shot) is broadcasting done from a location away from a formal television studio and is considered an electronic field production (EFP). A remote pickup unit (RPU) is usually used to transmit the audio and/or video back to the television station, where it joins the normal airchain. Other methods include satellite trucks, production trucks and even regular telephone lines if necessary.

History

The first airing of a remote broadcast came in 1924, when Loew's Theater publicist and WHN (New York City) station manager Nils Granlund leased telegraph lines from Western Union to provide the first link in what became called cabaret broadcasting." By early 1925, Granlund had established remote lines between WHN and more than thirty New York City jazz nightclubs, including the Silver Slipper, The Parody Club, the Cotton Club, the Strand Roof, and Club Moritz. These big band remotes would become a staple of the old-time radio era, lasting well into the 1950s.

Nils T. Granlund cited the 1925 WHN airing of Senator James J. Walker's announcement of his New York City mayoral candidacy through a remote broadcast from the New York Press Club as the first such remote link for a political forum.

In Latin America, on October 27, 1920 Dr Sussini made the first remote transmission in Argentina from the theatre El Coliseo in Buenos Aires. In Mexico on September 27, 1921, Adolfo Gomez Fernandez made a transmission from the Teatro Ideal, Mexico DF

The very first live remote broadcast to the nation was by the Canadian Broadcasting Corporation in 1938 when Frank Willis reported on the Moose River Gold Mine disaster in Nova Scotia http://archives.cbc.ca/economy_business/natural_resources/clips/3860/

On June 11, 1955, NBC, The National Broadcasting Company, provided the 1st live remote broadcast to the nation from Niagara Falls, New York.

Radio

In radio, remotes are often used for special events, such as concerts or sporting events, where either the entire event or advertisements for the event are broadcast on location. The cost of personnel and equipment is usually paid for by the host at each performance. However, if the event is recurring, such as a weekly broadcast from a nightclub, then dedicated lines are usually installed by the local telephone company in order to save on costs. With low range radio stations, and at events with no telephone lines, several radio stations call into the studio request line with a cell phone and microphone setup. From there, another dj in the studio will put them on-location dj live on the air via the studio request line. Some stations use this method when doing live broadcasts in areas where signal is weak.

Originally, analog audio broadcasts were sent through telephone hybrids, which, although low quality, were found to be acceptable for voice broadcasts. Later, frequency extenders were developed that used additional lines, shifting higher treble audio frequencies down on one end and back up on the other, providing a reasonable reproduction of the original sound. Currently, digital lines, such as ISDN or DSL, are used to send compressed digital audio back to the studio. In addition, modern remote pickup units have become extremely portable and can transmit single-channel monophonic FM-quality audio over regular telephone lines using built-in modems and advanced compression algorithms (MPEG-4, etc.).

Television

In TV, live television remotes are an almost daily part of television news broadcasts in the U.S. As a part of electronic news gathering (ENG), remotes are meant to bring the audience to the scene of the action, although often nothing directly related to the story is happening there anymore. In some cases live remotes have even taken ridicule, particularly for reporters who stand out in the middle of a hurricane's winds, risking their lives while telling viewers not to do the same.

To get to the scene quickly, a live remote may be done from a helicopter.

Live television remotes may often be used in a manner similar to radio remotes (and vice versa) as well.

Microbroadcasting

Microbroadcasting is the process of broadcasting a message to a relatively small audience.

Microbroadcasting, in radio terms, is the use of low-power transmitters (often Title 47 CFR Part 15 in the United States, or its equivalent elsewhere) to broadcast a radio signal over the space of a neighborhood or small town. Similar to pirate radio broadcasting, microbroadcasters generally operate without a license from the local regulation body, but sacrifice range in favor of using legal power limits (for example, 100 mW for medium wave broadcasts in the United States). Higher power levels can be achieved using carrier current techniques, which are widely used in colleges and universities. Both AM and FM bands are used, although AM tends to have better propagation characteristics at low power.

Microbroadcasting is also used by schools and businesses to serve just the immediate campus of the operation; well-known uses include audio tour guide systems, airport information services, and drive-in theaters, which often provide movie audio over the driver's car audio system. It has also been adopted as an advertising technique, particularly by car dealers and real estate agents.

Broadcast syndication

In broadcasting, syndication is the licensing of the right to broadcast television and radio programs by multiple television and radio stations, without going through a broadcast network. It is common in countries where broadcast programming is scheduled by television networks with local independent affiliates, particularly in the United States. Syndication is less common in the rest of the world, as most countries have centralized networks or television stations without local affiliates and syndication, although shows can also be syndicated internationally. The three main types of syndication are "first-run syndication", which is programming that is broadcast for the first time as a syndicated show and is made specifically to sell directly into syndication; "off-network syndication", which is the licensing of a program that was originally run on network TV or in some cases first-run syndication (colloquially called a "rerun"); and public broadcasting syndication.

Types

- First-run syndication refers to programming that is broadcast for the first time as a syndicated show and is made specifically to sell directly into syndication (not any one particular network), or at least first so offered in a given country (programs originally created and broadcast outside the US, first presented on a network in their country of origin, have often been syndicated in the US and in some other countries);
- Off-network syndication involves the licensing of a program that was originally run on network television or in some cases first-run syndication: a rerun; these are usually found on stations affiliated with smaller networks like Fox or The CW, especially since these networks broadcast one less hour of prime time network programming than the Big Three television networks. This usually occurs when a show has built up about four seasons worth or a minimum of 88–100 episodes. Successful shows in syndication can cover production costs and make a profit, even if the first run of the show was not profitable.
- Public broadcasting syndication has arisen in the US as a parallel service to member stations of the Public Broadcasting Service (PBS) and the handful of independent public broadcasting stations. This form of syndication more closely resembles the news agency model, where nominally competing networks share resources and rebroadcast each other's programs. For example, National Public Radio (NPR) stations commonly air the Public Radio Exchange's *This American Life*, which may contain stories produced by NPR journalists.

When syndicating a show, the production company, or a distribution company called a syndicator, attempts to license the show to one station in each media market or area, or to a commonly owned station group, within the country and internationally. If successful, this can be lucrative, but the syndicator may only be able to license the show in a small percentage of the markets. Syndication differs from licensing the show to a television network. Once a network picks up a show, it is usually guaranteed to run on most or all the network's affiliates on the same day of the week and at the same time (in a given time zone, in countries where this is a concern). Some production companies create their shows and license them to networks at a loss, at least at first, hoping that the series will succeed and that eventual off-network syndication will turn a profit for the show. A syndicated program is licensed to stations for "cash" (the stations purchase the rights to insert some or all of the advertisements at their level); given to stations for access to airtime (wherein the syndicators get the advertising revenue); or the combination of both. The trade of program for airtime is called "barter".

In the US (as a result of continued relaxation of station ownership regulations since the 1970s), syndicated programs are usually licensed to stations on a group level, with multiple stations owned and/or operated by the same broadcasting group carrying the program in different markets (except in areas where another station holds the market rights to the program) – making it increasingly more efficient for syndicators to gain widespread national clearances for their programs. Many syndicated programs are traditionally sold first to one of five "key" station groups (ABC Owned Television Stations, NBC Owned Television Stations, CBS Television Stations, Fox Television Stations and Tribune Broadcasting), allowing their programs to gain clearances in the largest US TV markets (such as New York City, Los Angeles, Chicago and Philadelphia, where all five aforementioned groups each own stations), before striking deals with other major and smaller station owners. Shows airing in first-run syndication that are carried primarily by an owned-and-operated station group of a network may sometimes be incorrectly referenced as a network program, regardless to its distribution to stations of varying network affiliations and despite the fact it is not part of an individual network's base schedule.

Since the early 2000s, some programs being proposed for national distribution in first-run syndication have been test marketed on a selected number of or on all stations owned by certain major station groups, allowing the distributor to determine whether a national rollout is feasible based on the ratings accrued in the selected markets where the program is being aired.

While market penetration can vary widely and revenues can be unreliable, the producers often enjoy more content freedom in the absence of network standards and practices departments; frequently, some innovative ideas are explored by first-run syndicated programming which the networks are leery of giving airtime to. Meanwhile, top-rated syndicated shows in the United States usually have a domestic market reach as high as 98%. Very often, series that are aired in syndication have reduced running times. For example, a standard American sitcom runs 22 minutes, but in syndication it may be reduced to 20 minutes to make room for more commercials.

Syndication can take the form of either weekly or daily syndication. Game shows, some "tabloid" and entertainment news shows, and stripped talk shows are broadcast daily on weekdays, while most other first-run syndicated shows are broadcast on a weekly basis and are usually aired on weekends only. There was a big discussion sometime in the 1990s and 2000s about whether a show could be syndicated while it was still on the air. There had been much opposition to this idea and it was generally viewed as leading to a death of the show. However, licensing a program for syndication actually resulted in the increased popularity for shows that were still in production. A prime example is *Law & Order*.

First-run Syndication in the U.S.

As with radio in the U.S., television networks, particularly in their early years, did not offer a full day's worth of programming for their affiliates, even in the evening or "prime time" hours. Some stations were not affiliated with any network, operating as independent stations. Both groups sought to supplement their locally produced programming and whatever network feeds there were with content that could be flexibly scheduled. The development of videotape and, much later, enhanced satellite downlink access furthered these options. While most past first-run syndicated shows were shown only in syndication, some canceled network shows continued to be produced for first-run syndication or were revived for syndication several years after their original cancellation. Until about 1980, most syndicated series were distributed to stations either on 16mm film prints (off-network reruns, feature films, and cartoons) or videotape (topical series such as the talk shows of Mike Douglas and Merv Griffin, and variety and quiz shows.)

Ziv Television Programs, Inc., after establishing itself as a major radio syndicator, was the first major first-run television syndicator, creating several long-lived series in the 1950s and selling them directly to regional sponsors, who in turn sold the shows to local stations. Ziv's first major TV hit was *The Cisco Kid*. Ziv had the foresight to film the Cisco Kid series in color, even though color TV was yet to be a reality. Among the most famous and widely watched Ziv offerings were *Sea Hunt*, *I Led Three Lives* and *Highway Patrol*. Some first-run syndicated series were picked up by networks in the 1950s and early 1960s, notably *The Adventures of Superman* and *Mr. Ed*. The networks started syndicating their reruns in the late 1950s, and first-run syndication shrank sharply, for a decade. Some stalwart series continued, notably *Death Valley Days*; other ambitious projects were also to flourish, however briefly, such as *The Play of the Week* (1959–1961), produced by David Susskind (of the syndicated talk show *Open End* and also producer of such network fare as *NYPD*).

Other popular syndicated series of the 1950s included MCA's *The Abbott and Costello Show* (vaudeville-style comedy) and Guild Films' *Liberace* (musical variety) and *Life With Elizabeth*, a domestic situation comedy that introduced Betty White to a national audience. In addition to *The Adventures of Superman*, many other series were based on comic strips and aimed at the juvenile audience, including *Flash Gordon*, *Dick Tracy*, *Sheena, Queen of the Jungle*, and *Joe Palooka*. Original juvenile adventure series included *Captain Gallant of the Foreign Legion*, *Cowboy G-Men*, and *Ramar of the Jungle*. Series based on literary properties included *Sherlock Holmes*, *Long John Silver* (based on *Treasure Island*), *The Invisible man*, and *The Three Musketeers*. Several of these were co-productions between US and European (usually British) companies. *Crusader Rabbit* pioneered in the area of first-run animated series; followed by *Bucky and Pepito*, *Spunky and Tadpole*, *Q.T. Hush*, and others. (All of these were five-minute shorts designed to be placed within locally hosted kiddie shows.) Syndicated sports programming included *Championship Bowling* and *All-Star Golf*, both produced by Chicago-based Walter Schwimmer, Inc.

In addition to regular series, syndicators also offered packages of feature films, cartoons, and short subjects originally made for movie theatres. Until late in the 1950s, however, much of the theatrical product available consisted of low-budget secondary features (mainly Westerns) with relatively few notable stars. One syndication company, National Telefilm Associates, attempted to create a "film network" of stations showing its lineup of first-run series, which included syndicated programs such as *Police Call* (1955), *How to Marry a Millionaire* (1957-1959), *The Passerby*, *Man Without a Gun* (1957-1959), and *This is Alice* (1958). The venture lasted several years, but was ultimately unsuccessful.

However, FCC rulings in the late 1960s curtailed the U.S. networks' ability to schedule programming in what has become known as the "early fringe", notably the 7–8 p.m. (Eastern and Pacific

Time) hour of "prime time", with the stated hope that this might encourage more local programming of social and cultural relevance to communities (off-network syndicated repeats were also banned); some projects of this sort came to fruition, though usually relatively commercial and slick ones such as Group W's *Evening Magazine/PM Magazine* franchise, and such pre-existing national projects as the brief commercial-television run of William F. Buckley, Jr.'s interview/debate series *Firing Line*. The more obvious result was an increase in Canadian-produced syndicated dramatic series, such as *Dusty's Trail* and the Colgate-sponsored *Dr. Simon Locke*. Game shows, often evening editions of network afternoon series, flourished, and a few odd items such as *Wild Kingdom*, canceled by NBC in 1971, had a continuing life as syndicated programming tailor-made for the early fringe.

1970s and 1980s

In 1971, the U.S. Federal Communications Commission passed the Prime Time Access Rule and Financial Interest and Syndication Rules, which prevented networks from programming one particular hour of prime time programming on its television stations each night and required the networks to spin off their syndication arms as independent companies. Although the intent of the rule was to encourage local stations to produce their own programs for this time slot, budgetary limits instead prompted stations to buy syndicated programs to fill the slot. This, coupled with an increase in UHF independent stations, caused a boom in the syndication market. In the 1970s, first-run syndication continued to be an odd mix: cheaply produced, but not always poor quality, "filler" programming. These included the dance-music show *Soul Train*, and 20th Century Fox's *That's Hollywood*, a television variation on the popular *That's Entertainment!* theatrically released collections of film clips from the Metro-Goldwyn-Mayer library.

There were also many imported programs distributed this way. These include the documentary series *Wild, Wild World of Animals* (repackaged by Time-Life with narration by William Conrad) and Thames Television's sober and necessarily grim *The World at War*. *The Starlost* (1973) was a Canadian series, apparently modified from the vision of science fiction writers Harlan Ellison and Ben Bova. Britain's ITC Entertainment, headed by Lew Grade, made *UFO* (1970) and *Space: 1999* (1975). These two series were created by Gerry Anderson (and his associates), who was previously best known for supermarionation (a combination of puppetry and animation) series such as *Thunderbirds*. The most successful syndicated show in the United States in the 1970s was probably *The Muppet Show*, also from Lew Grade's company. Animated series from the 1980s *Dogtanian and the Three Muskehounds* and *Around the World with Willy Fog* came from Spanish animation production company BRB Internacional and their Japanese co-producers Nippon Animation.

Game shows thrived in syndication during the decade. Nightly versions of *What's My Line?*, *Truth or Consequences*, *Beat the Clock* and *To Tell the Truth* premiered in the late 1960s and found loyal audiences for many years. Several daytime network games began producing once-a-week nighttime versions for broadcast in the early evening hours, usually with bigger prizes and often featuring different hosts (emcees were limited to appearing on one network and one syndicated game simultaneously) and modified titles (*Match Game PM*, *The $100,000 Name That Tune* or *The $25,000 Pyramid*, for example). A few independent game shows, such as *Sports Challenge* and *Celebrity Bowling*, also entered the syndication market around this time. Of these shows, *Let's Make a Deal* and *Hollywood Squares* were the first to jump to twice-a-week syndicated versions around 1973. Another popular daytime show to have a weekly syndicated version was *The Price Is Right*, which began concurrently in weekly syndication and on CBS; the syndicated "nighttime" version was hosted by Dennis James for its first five years, after which daytime host Bob Barker took over for another three years of weekly episodes (even though, by this point, the daytime and nighttime shows had diverged noticeably). The nighttime version of *Family Feud* (1977) quickly jumped from once-weekly to twice, and finally to five-day-a-week airings, and its massive popularity, along with that of new five-day-a-week entries like Jack Barry's *The Joker's Wild* (1977) and *Tic-Tac-Dough* (1978) and Chuck Barris's increasingly raunchy remakes of his 1960s hits *The Newlywed Game* and *The Dating Game*, brought an end (with rare exceptions) to the era of once-a-week games. Also popular in first-run syndication and daytime was *The Gong Show*, hosted by Barris throughout most of its run (Gary Owens hosted the first syndicated season).

A number of half-hour musical-variety shows were also offered in the early 1970s, generally built around personable middle-of-the-road singers like Bobby Vinton, Bobby Goldsboro, Dolly Parton, and Andy Williams, or groups like Sha Na Na, The Johnny Mann Singers, and The Golddiggers. *Wait Till Your Father Gets Home* (1972) was a Hanna-Barbera cartoon series attempting to ape the *All in the Family*-style sitcoms; *Skippy the Bush Kangaroo* (1969) was an Australian children's series in the manner of *Flipper* or *Gentle Ben* (a decade later, the decidedly not-for-children Australian *Prisoner: Cell Block H* would have a brief U.S. syndicated run); and a Canadian sketch-comedy series began appearing on U.S. television stations in 1977 – *Second City Television*, which would eventually find a home, for two seasons, on NBC, as *SCTV Network 90* (and on premium cable channel Cinemax by 1983).

The Universal/Paramount-produced package of original programming, *Operation Prime Time*, began appearing on *ad hoc* quasi-networks of (almost by necessity) non-network stations in the U.S. in 1978, with a mini-series adaptation of John Jakes's *The Bastard*. From the later 1960s into the late 1970s, Westinghouse also found considerable success with *The Mike Douglas Show*, a variety/talk show hosted by a singer with an easygoing interview style, which aired in the afternoons in most markets; similar programs soon followed featuring Merv Griffin, who had been the host of CBS' most sustained late-night answer to *The Tonight Show Starring Johnny Carson* previously, and another network veteran, Dinah Shore. Also notable was the growing success of audience-participation talk shows, particularly that of the innovator of the format, Phil Donahue.

First-run syndication in the 1970s also made it possible for some shows that were no longer wanted by television networks to remain on the air. In 1971, ABC canceled *The Lawrence Welk Show*, which went on to produce new episodes in syndication for another 11 years, and currently continues to much success in weekend reruns (with new segments featuring Welk cast members inserted within the episodes) distributed to PBS stations by the Oklahoma Educational Television Authority. Also in 1971, CBS dropped *Lassie* and *Hee Haw*, the latter show's run ending as part of the network's cancellation of all of its rural-oriented shows (known then as "rural purge", which also resulted in the cancellations of *The Beverly Hillbillies* and *Green Acres*). *Lassie* entered first-run syndication for two years, while *Hee Haw* continued to produce new episodes until 1992.

First-run Syndicated Comedy

Throughout the mid-to-late 1980s into the early 1990s, sitcoms continued to enter first-run syndication after being canceled by the networks, the most successful of which were *Mama's Family* and *Charles In Charge*. Other sitcoms during this time to enter first-run syndication after network cancellation included *Silver Spoons, Punky Brewster, Webster, It's a Living, Too Close for Comfort, 9 to 5, What's Happening!!*, and *WKRP in Cincinnati*. Many of these sitcoms produced new shows in syndication mainly to have enough episodes for a profitable run in reruns. Other sitcoms, such as *Small Wonder, Out of This World, The Munsters Today*, and *Harry and the Hendersons* (as well as more action-adventure oriented series like *Superboy* and *My Secret Identity*) enjoyed success in syndication throughout their entire run.

Dramatic First-run Syndicated Programs

The broadcast networks aired many action-adventure programs from the 1950s to the 1980s. By the late 1980s, however, increasing production costs made them less attractive to the networks. Studios found that reruns of one-hour dramas did not sell as well as sitcoms, so they were unable to fully recoup the shows' costs using the traditional deficit financing model. Such shows instead moved to syndication, starting with *Star Trek: The Next Generation*, which debuted in 1987 and became one of the most watched syndicated shows throughout its seven-year run. Its great success caused many others to debut; by 1994, there were more than 20 one-hour syndicated shows.

When NBC canceled the television series adaptation of *Fame* after only two seasons, the producers made special arrangements with LBS Communications, which resulted in MGM reviving the series for first-run syndication in the fall of 1983, where it continued for four more seasons, with the last first-run episode airing in the U.S. on May 18, 1987. *Friday the 13th: The Series* (a horror series which shared its title with the successful movie franchise) also debuted in 1987. The next syndicated shows that debuted in 1988 were *War of the Worlds* and *Freddy's Nightmares*.

Baywatch, which debuted in 1989 on NBC and was canceled after one season also became one of the most watched syndicated shows throughout its ten-year-run, garnering a worldwide audience. Another series to be one of the few shows to nearly match *Baywatch* would be the television series *Renegade*. *Star Trek: Deep Space Nine* was also syndicated. *Hercules: The Legendary Journeys* and its spin-off series *Xena: Warrior Princess* helped build the audiences for such shows; *Babylon 5* and *Forever Knight* drew devoted "cult" audiences; *Psi Factor* and *Poltergeist: The Legacy* attempted to draw on the audience for the Fox series *The X-Files* (as did, even less probably, the short-lived spinoff *Baywatch Nights*).

Among the other series were *Relic Hunter, VIP, High Tide, She Spies* and *Once a Thief.* In 1997, *Earth: Final Conflict*, based on ideas from Gene Roddenberry, premiered in syndication. Three years later, a second Gene Roddenberry series, *Andromeda* also premiered in syndication. After that series' ended in 2004, there would not be a first-run syndicated drama (or a first-run scripted series in syndication) until 2008, when Disney-ABC Domestic Television and ABC Studios teamed up with Sam Raimi to launch a new first-run syndicated series *Legend of the Seeker*, based on Terry Goodkind's *Sword of Truth* novel series. Another gap in first-run scripted series in syndication followed four years after *Legend of the Seeker* was canceled in 2009, until Trifecta Entertainment & Media (a company that mainly distributes programs for off-network syndication) began producing *SAF3* in 2013.

Animated Series

During the late 1970s and 1980s, independent stations signed on in mid-sized and many small markets. The market for made-for-television cartoons grew as a result to include a branch for such stations. It usually had a greater artistic freedom, and looser standards (not mandated by a network). The older Bugs Bunny and Popeye cartoons made way for first-run syndicated cartoons such as *He-Man and the Masters of the Universe, Rambo: The Force of Freedom, ThunderCats, Dennis the Menace, My Little Pony, The Transformers, G.I. Joe, Voltron, M.A.S.K., The Centurions, Teenage Mutant Ninja Turtles*, and reruns of *Scooby-Doo, Garfield and Friends*, and *The Pink Panther*, among many others.

In 1987, The Walt Disney Company tried its luck at syndication; *DuckTales* premiered that September and would eventually last for 100 episodes. The success of *DuckTales* paved the way for a second series two years later, *Chip 'n Dale Rescue Rangers*. The following year, the two shows aired together under the umbrella block *The Disney Afternoon*. In fall of 1990, Disney added another hour to *The Disney Afternoon*; the block continued in syndication, running additional first-run animated series until 1999.

These cartoons initially competed with the ones that were nationally televised on the broadcast networks. In the 1980s, national broadcast networks only aired cartoons on Saturday mornings, not competing with the weekday and Sunday syndication blocks aired by local independent stations; however, by the 1990s, Fox and then The WB launched their own weekday afternoon children's program blocks. By the end of the 1990s, both syndication distributors and broadcast networks ended up losing most of its children's market to the rise of cable television channels aimed at that audience such as Nickelodeon and Cartoon Network, which provided appealing children's entertainment throughout the week at nearly all hours.

Syndication remains a method of choice for distributing children's programming, although this has gradually shifted to only produce programs to satisfy the federally mandated "educational-informational" (E/I) rule imposed in the late 1990s as part of an amendment to the Children's Television Act of 1990 that requires stations to air three hours of educational children's programs every week, regardless of the station's format. Syndication is generally a less expensive option for a local station than to attempt to produce its own locally originated E/I programming; not all networks provide their own E/I programs, so stations that are affiliated with networks that do not carry children's program blocks acquire E/I programs off the syndication market to fulfill the requirements.

News Programming and Late-night Talk Shows

Also in the 1980s, news programming of various sorts began to be offered widely to stations. *Independent Network News*, which was produced by WPIX in New York City, was a half-hour nightly program that ran from 1980 to 1990 on independent stations (in some markets, *INN* was paired with a locally produced primetime newscast); CNN would offer a simulcast of programming from its sister network Headline News (now HLN) to broadcast stations later, though it was used mainly to fill overnight time periods and was effectively discontinued in syndication with the 2006 launch of that network's "Headline Prime" talk show block. *Entertainment Tonight* began its long and continuing run as a "soft" news daily strip, with a number of imitations following (among which have included such entertainment news shows as *TMZ on TV*, *Extra* and *ET*'s own spin-off *The Insider*); and "tabloid" television, in the wake of ABC's *20/20* and, more immediately, 20th Television's *A Current Affair*, would become a syndication staple with such series as *Hard Copy* and *Real TV*.

Another area where network dominance was challenged by syndicated programming in the 1980s was with late-night talk shows; *The Arsenio Hall Show* was the only very successful one (it would be canceled after five years in 1994 due to ratings declines spurred by many CBS affiliates pushing the show to later timeslots following the debut of the *Late Show with David Letterman*, and was later revived in 2013), but similar programs were attempted such as Alan Thicke's earlier short-lived *Thicke of the Night*, Lauren Hutton's innovatively shot *Lauren Hutton and...*, and talk shows hosted by Dennis Miller, Whoopi Goldberg, David Brenner and Keenen Ivory Wayans; Magic Johnson's *The Magic Hour* was seen as a massive flop, similar to *Thicke of the Night*. The popularity of syndicated talk shows fell dramatically in the mid-1990s as network and cable offerings expanded in the wake of Johnny Carson's retirement.

Reality and Live-action Children's Shows

Long before their popularity on network television from the 2000s onward, reality competition shows in one form or another, such as *Star Search* and *American Gladiators*, enjoyed popularity in syndication as early as the mid-1980s. Since the now-defunct networks UPN and The WB began offering their affiliates additional nights of prime time programming in the late 1990s, there have been fewer first-run scripted series in syndication, at least, in the U.S.; much as with the closing of windows that provided opportunity for Ziv in the 1950s and various producers in the early 1970s. The more expensive dramatic projects are less attractive to syndicators (particularly when they might be sold, with somewhat less risk, to cable channels); "reality" series such as *Cheaters* and *Maximum Exposure* and several dating series began to be more common in the early 2000s. Some of the more low-key programs in this category were designed to appeal to children, such as *Beakman's World*, *Animal Rescue* and *Jack Hanna's Animal Adventures*. They were able to get significant clearance because of stricter Federal Communications Commission (FCC) enforcement of rules on children's television programming.

Game Shows

Several game shows are currently syndicated; historically, the most popular had been *Wheel of Fortune* and the current version of *Jeopardy!*, both created by television personality Merv Griffin, respectively premiering in 1983 and 1984. The shows have been #1 and #2 or #1 to #3 in the syndication ratings consistently since at least the late 1980s. In fact, according to the *Guinness Book of World Records*, *Wheel* is the most popular syndicated television program both within and outside the United States. *Family Feud* ended its first syndication run in 1985. Three years later, a revival of the program became a moderate hit and continued for seven seasons. A third revival hit the airwaves in 1999 and has gone through four hosts. The first three hosts (Louie Anderson, Richard Karn and John O'Hurley) struggled in their respective runs and only lasted three to four years. The current run of the program, hosted by Steve Harvey, has been a major ratings success; on the week of June 12, 2015, for the first time ever, *Family Feud* was the highest-rated syndicated program in terms of average household ratings. A *Hollywood Squares* revival also thrived beginning in 1998, running six seasons until its 2004 cancellation. By far the most successful entry into the market in the 2000s has been the daily version of *Who Wants to Be a Millionaire*, which premiered in 2002.

Because game shows are very inexpensive to produce, with many episodes completed each day of production, successful ones are very profitable; for example, in 1988 *Jeopardy!* cost an estimated

$5 million to produce but earned almost $50 million in revenue. New game show concepts (that is, not based on an existing or pre-existing format) are rarely tried and usually unsuccessful in syndication; somewhat of an exception to this was *Street Smarts*, which lasted from 2001 to 2006 (despite the series airing in late night slots in many markets). Between 2003 and 2007, no new game shows debuted in syndication, marking four consecutive seasons where no new shows with that genre debuted, a syndication first. That streak ended with the fall 2007 debuts of *Temptation* and *Merv Griffin's Crosswords*, bringing the daytime tally to six game shows; both ended production after one year, though *Crosswords* aired in reruns in some cities during the 2008–09 season before those reruns moved exclusively to cable.

More new shows were added for the 2008–09 fall season, including a daytime run of *Deal or No Deal* (which featured certain elements that differed from the show's franchised format, most notably with prospective players instead of models holding briefcases that held the monetary amounts) and an adaptation of the popular board game *Trivial Pursuit*. While *Deal* caught on and was renewed for the 2009–2010 season, *Trivial Pursuit: America Plays* suffered low ratings throughout its run and was canceled.

For the 2009–2010 season, the Fox game show *Are You Smarter than a 5th Grader?* moved to syndication with a new, less expensive format. *Don't Forget the Lyrics!* followed for the 2010–2011 season. *Deal*, suffering from falling ratings, was canceled in February 2010, with the final episodes airing in late May of that same year. *5th Grader* and *Don't Forget the Lyrics* were canceled the following year for the same reason (although reruns of *5th Grader* remain on cable). Reruns of the popular Discovery Channel show *Cash Cab* began airing in syndication in January 2011. Reruns of the GSN dating game show *Baggage* first aired in syndication as a test run in early 2011 on stations owned by the Sinclair Broadcast Group, which preceded its full launch into other markets in fall 2012; although it was removed from syndication after one season.

The 2014-15 season saw the introduction of *Celebrity Name Game*, hosted by former *The Late Late Show* host Craig Ferguson; the series was renewed for a second season in January 2015, while Ferguson would also win a Daytime Emmy Award for Outstanding Game Show Host for his work on the program. In January 2016, Fox owned-and-operated stations began a test run of *South of Wilshire*—a game show produced by TMZ.

Stripped Talk Shows

The dominant form of first-run syndication in the U.S. for the last three decades has been the "stripped" (or "strip") talk show, such as *Donahue*, *Oprah Winfrey*, *The Tyra Banks Show*, and *The Jerry Springer Show*. Strip programming is a technique used for scheduling television and radio programming to ensure consistency and coherency. Strip programming is used to deliver consistent content to targeted audiences. Broadcasters know or predict the times at which certain demographics will be listening to or watching their programs and play them at that time. As with game shows, talk shows are inexpensive to produce and very profitable if successful. In many markets, a stripped show will be seen twice daily, usually with different episodes (one being a more recent episode and the other being an episode from a previous season). Sometimes, station groups with more than one station in a market, or a "duopoly", will run one episode of a strip on one of their stations in the morning, and the other available episode on another of their stations that night.

Meanwhile, the popularity of some of the audience-participation talk shows continues to encourage new participants, some of whom, such as Morton Downey, Jr. and Rosie O'Donnell, have brief periods of impressive ratings and influence; others, such as Oprah Winfrey and Maury Povich, have a sustained run. A notable scheduling decision was made by KRON-TV in San Francisco: a 2000 dispute with NBC led to that station's disaffiliation from that network after 52 years, and since all the other larger networks were already represented in San Francisco, KRON decided to become one of the largest commercial independent stations by market size on the VHF band in the U.S., and soon tried running *Dr. Phil*, a popular new stripped series hosted by Winfrey-associate Phil McGraw, in primetime, with impressive ratings results.

2000s

First-run syndicated shows in the US include talk shows (e.g., *The Dr. Oz Show*, *Dr. Phil* and *The Ellen DeGeneres Show*); tabloid/newsmagazine shows (e.g., *TMZ Live*); crime/law enforcement shows (e.g., *Crime Watch Daily*); game shows (e.g., *Family Feud*, *Jeopardy!* and *Wheel of Fortune*); reality court shows (e.g., *Judge Judy*); and sitcoms (e.g., *The First Family*).

Influence on Television Schedules

In earlier times, independent stations thrived on syndicated programming, including some venerable and quite profitable stations such as KMSP-TV in the Minneapolis-St. Paul market. With the loosening of FCC regulations and the creation of new additional broadcast networks (such as Fox, The CW, MyNetworkTV and Ion Television), most of these independents have joined one or another of these or smaller (religious or low-budget) networks.

In another case, like those of KCAL-TV in Los Angeles, KMCI-TV in Lawrence-Kansas City and WMLW-TV in Racine-Milwaukee, those independent stations are used to complement their network-affiliated sister station (respectively in the mentioned cases, KCBS-TV, KSHB-TV and WDJT-TV) by allowing a duopoly control of more syndicated programming than would be possible on one station (and to spread it throughout the schedule of the two stations, often several times a day), or to air news programming in times unavailable on the larger network station, along with fulfilling network and syndicated programming commitments, which allows popular or network programming to be moved to the independent stations due to breaking news or sports commitments without the traditional inconvenience of a late night or weekend airing of the pre-empted show. A duopoly of a network-affiliated and independent station also allows a network station to move a low-rated syndicated program to their sister independent station to stem revenue losses.

Off-network Syndication

Off-network syndication occurs when a network television show is syndicated in packages containing some or all episodes, and sold to as many television stations and markets as possible to be used in local programming timeslots. In this manner, sitcoms are preferred and more successful because they are less serialized, and can be run non-sequentially, which is more beneficial and less costly for the station. In the United States, local stations now rarely broadcast reruns of primetime dramas (or simply air them primarily on weekends); instead, they usually air on basic cable channels, which may air each episode 30 to 60 times.

Syndication rights typically last for six consecutive showings of a series within three to five years; if a program continues to perform well enough in broadcast or cable syndication during the initial cycle, television stations or cable networks can opt to renew an off-network program for an additional cycle.

Syndication has been known to spur the popularity of a series that only experienced moderate success during its original network run. The most notable example of this is *Star Trek*, which ran for three seasons on NBC from 1966 to 1969, garnering only modest ratings, but became a worldwide cult phenomenon after it entered off-network syndication. Its success in syndication led to the *Star Trek* film series, *Star Trek: The Next Generation*, and several other series.

It is common for long-running series to have early seasons syndicated while the series itself is still in first-run network production. In order to differentiate between new and rebroadcast content, up until the 1980s it was not uncommon for series to be syndicated under a different title than that used in their original broadcast run. Examples include *Bonanza* (which was syndicated as *Ponderosa*), *Gunsmoke* (as *Marshall Dillon*), *Emergency!* (as *Emergency One*), *Ironside* (as *The Raymond Burr Show*), *Hawaii Five-O* (as *McGarrett*), *M*A*S*H* (as *M*A*S*H 4077th*), *Marcus Welby, M.D.* (as *Robert Young, Family Doctor*), *CHiPs* (as *CHiPs Patrol*), and *Happy Days* (as *Happy Days Again*).

Syndication of older episodes can also increase exposure for a television show that is still airing first-run network episodes. In the case of the CBS sitcom *The Big Bang Theory*, its syndication, particularly on TBS, is one of the reasons attributed for a rise in first-run ratings for its sixth season. The sixth season episode "The Bakersfield Expedition", for example, was the first episode of that series to attract 20 million viewers.

Strip/Daily Syndication

Off-network syndication can take several forms. The most common form is known as strip syndication or daily syndication, when episodes of a television series are shown daily five times a week in the same time slot. In the 1960s and 1970s, independent stations with no news departments began viewing strip syndication as a necessary means of obtaining effective counterprogramming to the local news programs airing on network affiliates. Typically, this means that enough episodes must exist (88 episodes, or four seasons, is the usual minimum, though many syndicators prefer a fully rounded 100 episodes) to allow for continual strip syndication to take place over the course of several months, without episodes being repeated. However, there are exceptions, such as the 65-episode block (common in children's programming), which allows for a 13-week cycle of daily showings, so there will only be four repeats in a year.

In some cases, more than one episode is shown daily. Half-hour sitcoms are sometimes syndicated in groups of two or four episodes, taking up one or two hours of broadcast time. If a series is not strip syndicated, it may be aired once a week, instead of five times a week. This allows shows with fewer episodes to last long in syndication, but it also may mean viewers will tire of waiting a week for the next episode of a show they have already seen and stop watching. More often, hour-long dramas in their first several runs in syndication are offered weekly; sitcoms are more likely to get stripped. In recent years, there has been something of a trend toward showing two consecutive episodes of a program on Saturday and Sunday nights after prime time (generally following the local news). This pattern has been particularly prominent for shows which are still in production but have run long enough to have many previous episodes available.

As with commercial stations, not all the airtime nor all the perceived audience are met by the productions offered U.S. public-broadcasting stations by PBS; additionally, there are some independent public television stations in the U.S. which take no programming from that (somewhat) decentralized network. As a result, there are several syndicators of programming for the non-profit stations, several of which are descendants of the regional station groups which combined some, not all, of their functions into the creation of PBS in 1969. American Public Television (APT) is the largest of these, nearly matched by the National Educational Telecommunications Association (NETA); similarly, the recently defunct Continental Program Marketing was another of the syndicator-descendants (of the Northeastern, Southeastern, and Rocky Mountain educational networks, respectively) of the pre-PBS era. Among the other notable organizations in the U.S. are Westlink Satellite Operations (based at Albuquerque's KNME), BBC Worldwide Americas (which often works with other distributors and individual stations, since it has no satellite access of its own in the U.S.), Deutsche Welle, Executive Program Services, the Program Resource Group and PBS's Long Island, New York member station WLIW, which is (with the arguable exception of KNME) the most prolific contributor of any individual station of syndicated programming, most obviously the *BBC World News*, *Doctor Who* and *Monty Python's Flying Circus* in the U.S.

Monetary Rates

In 1993, Universal Studios became one of the first studios to cash in on the cable trend, first selling repeats of *Major Dad* to USA Network in 1993 for $600,000 per episode, the first time a network program was exclusively sold to a cable network for its first run rights. Later it sold reruns of *Law & Order* to A&E for about $155,000 an episode; in 1996, the studio got $275,000 from USA Network for repeats of *New York Undercover*, a far less successful show. *Law & Order* drew A&E's highest daytime ratings – one million viewers per episode.

Universal sold reruns of *Xena: Warrior Princess* and *Hercules: The Legendary Journeys* to USA for $300,000 each. And even long-forgotten shows can find new life: The Nashville Network bought *The Dukes of Hazzard* from Warner Bros. in 1997 for well over $10 million. USA Network paid $750,000 for the rights to *Walker, Texas Ranger*; while USA's reruns of the show drew an average of 2.3 million viewers – outstanding by cable standards – Perth says the show will need *an enormous number of airings to have any sort of profitability."*

Year sold	Show	Studio	Cable network	Price*
1986	*Falcon Crest*	Lorimar-Telepictures	Turner Broadcasting	$10,000
	Knots Landing	Lorimar-Telepictures	Turner Broadcasting	$12,000
1988	*Murder, She Wrote*	Universal Television	USA Network	$525,000
1991	*Unsolved Mysteries*	Buena Vista Television	Lifetime	$180,000
1993	*The Commish*	ABC Productions	Lifetime	$195,000
1994	*Law & Order*	Universal Television	A&E	$155,000

1995	*Melrose Place*	Spelling Television	E!	$200,000
	Picket Fences	20th Television	FX	$190,000
	Lois & Clark: The New Adventures of Superman	Warner Bros.	TNT	$275,000
	Dr. Quinn, Medicine Woman	MTM Enterprises	The Family Channel	$250,000
	NYPD Blue	20th Television	FX	$400,000
1996	*Xena: Warrior Princess*	Universal Television	USA	$300,000
	Hercules: The Legendary Journeys	Universal Television	USA	$300,000
	Chicago Hope	20th Television	Lifetime	$475,000
	Homicide: Life on the Street	NBC Productions	Lifetime	$425,000
	The X-Files	20th Television	FX	$600,000
	Walker, Texas Ranger	Columbia TriStar	USA	$750,000
	ER	Warner Bros.	TNT	$1.2 million

Not all programs in syndication are sold for a fee. Less popular programming may be distributed by *barter*, in which the syndicator, instead of selling the show to a station, offers the show for free, with the caveat that the station give up its advertising time on other shows to the syndicator's advertisers. Barter syndication, in addition to the cost advantage, is popular because of its flexibility; a station can typically pick up a barter syndicated program for only a few weeks or months, without the long-term financial commitment of a traditional syndicated series, allowing the station to plug the show into its lineup to fill a hole in the schedule.

Types of Deals

Cash deals are when a distributor offers a syndicated program to the highest bidder. A cash plus deal is when the distributor retains advertising space to offset some of the cost for the program. The station gets the program for a little less in exchange for some ad space for the producer.

Barter deals are usually for new untested shows or older shows. In this type of deal, distributors get a fraction of the advertisement revenue in exchange for their program. For example, in a 7/5 deal the producer gets seven minutes of advertising time, leaving five minutes for the station to insert local as well as national advertisements.

Radio Syndication

Radio syndication generally works the same way as television syndication, except that radio stations usually are not organized into strict affiliate-only networks. Radio networks generally are only distributors of radio shows, and individual stations (though often owned by large conglomerates) decide which shows to carry from a wide variety of networks and independent radio providers. As a result, radio networks such as WestwoodOne or Premiere Networks, despite their influence in broadcasting, are not as recognized among the general public as television networks like CBS or ABC (many of these distributors ally themselves with television networks; WestwoodOne, for instance, is allied with NBC, while Premiere is allied with Fox). Some examples of widely syndicated commercial broadcasting music programs include weekly countdowns like Rick Dees' *Weekly Top 40*, the *American Top 40*, *American Country Countdown* with Kix Brooks, *Canada's*

Top 20 Countdown, the *Canadian Hit 30 Countdown* and the nightly program, *Delilah*, heard on many U.S. stations.

Syndication is particularly popular in talk radio. While syndicated music shows (with the exception of some evening and overnight shows such as *Delilah* mentioned above) tend to air once a week and are mostly recorded, most popular talk radio programs are syndicated daily and are broadcast live. Also, with relatively few 24-hour live talk radio networks (though this, in recent times, has been changing), most radio stations are free to assemble their own lineup of talk show hosts as they so choose. Examples of syndicated talk programs are Premiere Networks' *The Bob & Tom Show*, Dial Global's *The Jim Bohannon Show*, and the self-syndicated *The Dave Ramsey Show* (more recently, talk networks such as Talk Radio Network have been marketing and packaging all-day lineups, marking a departure from the syndication model; as such, popular shows such as Cumulus Media Networks' *The Savage Nation* and Premiere's *The Rush Limbaugh Show* now air as part of a broader network lineup in many markets, particularly on Premiere owned-and-operated stations, though they continue to be syndicated to non-network stations as well). Talk syndication tends to be more prevalent because voice tracking, a practice used by many music stations to have disc jockeys host multiple supposedly local shows at once, is not feasible with live talk radio.

National Public Radio, Public Radio International, and American Public Media all sell programming to local member stations in the U.S., most of which are subsidized through the Corporation for Public Broadcasting but operated by private nonprofit organizations, universities, state or local governments. This is in contrast to centralized public radio networks in other countries (such as Canada's CBC and the United Kingdom's BBC) that own and operate all of their stations as arms of the national government and run them as a strict network (from 1948 to 2013, the United States had a strict anti-propaganda law, the Smith–Mundt Act, that prohibited broadcasting government-owned networks such as Voice of America to American audiences. The law was mostly repealed in 2013, but distribution of VOA or other federally produced radio programming is still rare). Two independently produced, non-commercial syndicated programs, heard on hundreds of community radio and indie radio stations, are *Alternative Radio* and *Democracy Now!*. Some (in fact, most) radio programs are also offered on a barter system usually at no charge to the radio station. The system is used for live programming or preproduced programs and include a mixture of ad time sold by the program producer as well as time set aside for the radio station to sell.

History

Before radio networks matured in the United States, some early radio shows were reproduced on transcription disks and mailed to individual stations. An example of syndication using this method was RadiOzark Enterprises, Inc. based in Springfield, Missouri, co-owned with KWTO. The Assembly of God, with national headquarters in Springfield, sponsored a half-hour program on the station called *Sermons in Song*. RadiOzark began transcribing the show for other stations in the 1940s, and eventually 200 stations carried the program. The company later produced country music programs starring among others, Smiley Burnette, George Morgan, Bill Ring and Tennessee Ernie Ford (260 15-minute episodes of *The Tennessee Ernie Show* were distributed), and more than 1,200 U.S. and Canadian stations aired the programs. Many syndicated radio programs were distributed through the U.S. mail or another delivery service, although the medium changed as technology developed, going from transcription disks to phonograph records, tape recordings, cassette tapes and eventually CDs. Many smaller weekend programs still use this method to this day, though with the rise of the Internet, many stations have since opted to distribute programs via CD-quality MP3s through FTP downloads.

It was not until the advent of satellite communications in the 1980s that live syndication became popular (though it could be transmitted through network lines, it was not particularly common because of cost, network congestion and quality issues). Since then, most syndicated radio programs are distributed using satellite subcarrier audio technology. Shortly after satellite networks such as RKO, Transtar and SMN began, the Fairness Doctrine was repealed, which is credited with helping Rush Limbaugh become the first national talk radio superstar. At the same time, the FCC began issuing more FM radio licenses to suburban and rural areas in the late 1980s, which allowed more room for music stations on the FM dial; radio formats such as country music that were traditionally AM fixtures even after most pop and rock music moved to FM were now moving to FM as well, leaving much more room for talk formats on the AM dial. As the 1990s went on, Dr. Laura and Howard Stern began their national shows, rising to become national icons. The Telecommunications Act of 1996, which led to significant concentration of media ownership, facilitated the rapid deployment of both existing and new syndicated programs in the late 1990s, putting syndication on par with, and eventually surpassing, the network radio format.

After the September 11, 2001 terrorist attacks, syndicated talk radio saw a notably rapid rise in popularity, as networks rushed new national shows into syndication to meet the demand for discussion of national issues. Many of these, such as Laura Ingraham, Bill O'Reilly, Sean Hannity and Glenn Beck, were mostly supportive of the actions of the Republican-led government; a few others, such as Alex Jones, were openly critical of the government's actions and motives. After the Democrats took control in the late 2000s, the gap between the two styles narrowed due to the mutual opposition of both camps to the government's actions, which allowed Jones greater clearance on stations.

In contrast to conservative talk radio, which has predominantly been driven by syndication, progressive talk radio has almost always been a network-driven model. The incompatibility of conservative and progressive ideologies and the lack of syndicated progressive hosts required solutions that could produce all-day programming to individual stations. It was not until Air America Radio launched in 2004 that progressive talk would become viable; though it failed several years later, Dial Global now carries a network slate that is carried on most progressive talk stations. Sports radio is likewise mostly a network phenomenon, partially because the irregular nature of sports pre-emptions makes having a full-time network to be able to cut into and join in progress at any time highly convenient. Syndicated radio is not as popular in other parts of the world. Canada has a few independently syndicated shows, but the bulk of syndicated content there comes from the United States, and the sum total of syndicated programming is far less than most American stations, as Canadian stations rely more heavily on local content. Most other countries still follow the network radio model.

International Syndication

Syndication also applies to international markets. Same language countries often syndicate programs to each other – such as programs from the United Kingdom being syndicated to Australia and vice versa. Another example would be programs from the United Kingdom, Mexico, Brazil, and Argentina being syndicated to local television stations in the United States, and programs from the United States being syndicated elsewhere in the world. One of the best-known internationally syndicated television series has been *The Muppet Show*, which was produced by Grade's English ITV franchise company ATV at Elstree Studios in Hertfordshire, and was shown around the world, including the United States, where it aired in syndication (including the owned-and-operated stations of CBS), and Canada, where CBC Television aired the show. The 1970s was a time when many British comedies, including *The Benny Hill Show* and *Monty Python's Flying Circus* were syndicated to the United States and worldwide. Many soaps and long-running series are also successfully syndicated around the globe.

The hit television show *CSI* earned $1.6 million per episode in its first cycle in cable syndication. There were many different versions of the show making it an international success. It was already popular in the U.S., so becoming a success internationally as well as within the U.S. made syndication sensible. Whether a series is produced in the U.S. or not is based on the economic value and potential viability of its sales internationally with the possibility of syndication.

Economic factors that influence production outside the U.S. play a major role in deciding if a television show will be syndicated, internally and internationally. International syndication has sustained a growing of prosperity and monetary value amongst the distributors who sell to them. Due to a rise in competition, syndicators have upheld high standards for different countries to buy the rights to distribute shows. During the 1990s poor ratings were popular amongst syndicated shows, but distributors still made it possible for international competition to happen and buy U.S. shows. Colombian, Brazilian, Mexican and Venezuelan telenovelas are programmed throughout the Portuguese and Spanish-speaking world, and in many parts of India, China and Europe, while Turkish television drama is broadcast in the Balkans, some other European countries, Western and Central Asia and North Africa.

U.S.-style Syndication Internationally

Because of the structural differences discussed above, there are presently very few areas where a true U.S.-style syndication model operates, whereby programs are sold on a per-area basis (within a single country) to local or regional stations with differing (or no) network affiliations. Canada was historically one of the few exceptions. Until the mid-1990s, television stations in Canada, like those in the U.S., were typically run as separate local operations, with a small number of moderately sized ownership groups such as Baton, CanWest, WIC, and CHUM. Those stations that were affiliated with a national network, i.e. CBC or CTV, did not always receive a full schedule of programming from that network.

At this time, it was not uncommon for U.S. syndicators to treat Canada as an extension of their domestic syndication operations, for both their first-run and off-network offerings. This is still the case for American radio programs; Canadian radio networks are not assembled as rigidly into networks (except for the CBC's radio division). However, an alternate form of first-run syndication was performed by some domestic broadcasters: as the Canadian rights to U.S. primetime series were often acquired by individual station groups (as opposed to full-fledged national networks), they would in turn resell local rights for those programs to stations in areas where they did not operate.

Since the late 1990s, as most stations have been consolidated into national networks consisting almost entirely of owned-and-operated stations and with full-day network schedules, both types of syndication have largely disappeared from the Canadian broadcast landscape. Programs that are sold in syndication in the U.S. are now generally sold to Canadian media groups to air across all their properties, with per-market sales now being very rare (the minor exception being the few remaining independent stations that exist in Canada, such as CHCH-DT in Hamilton, Ontario). *The Oprah Winfrey Show* appears to have been the last significant holdout to this model (prior to its 2011 conclusion), having aired primarily on CTV stations, but in some markets airing instead on a Global station, and even some CBC affiliates.

Regional Syndication

There are three key reasons why a radio station will decide to pick up a syndicated show – the program is unique and difficult to replicate content, has a decent ratings track record, and whether or not it has a celebrity host. New developing radio programs are generally able to claim one of these attributes, but not all three. Regional syndication attempts to replace these benchmark attributes with other benefits that are generally recognized by the industry as also being important. Given the financial downturn within the industry, the need for quality cost effective locally relevant programming is greater than ever before. Programs that offer regionally specific content while providing the economic benefits of syndication can be especially appealing to potential affiliates. Regional syndication can also be more attractive to area advertisers who share a common regional trading area versus assembling a radio network of stations that hopscotch across the United States.

Broadcast Programming

Broadcast programming is the practice of organizing and or ordering of broadcast media programs (Internet, television, radio, etc.) in a daily, weekly, monthly, quarterly or season-long schedule.

Modern broadcasters use broadcast automation to regularly change the scheduling of their programs to build an audience for a new show, retain that audience, or compete with other broadcasters' programs. In the United Kingdom, this is known as TV listings.

Television scheduling strategies are employed to give programs the best possible chance of attracting and retaining an audience. They are used to deliver programs to audiences when they are most likely to want to watch them and deliver audiences to advertisers in the composition that makes their advertising most likely to be effective. Digitally based broadcast programming mechanisms are known as electronic program guides (EPG).

At a micro level, scheduling is the minute planning of the transmission; what to broadcast and when, ensuring an adequate or maximum utilization of airtime.

Programming History

With the beginning of scheduled television in 1936, television programming was initially only concerned with filling a few hours each evening – the hours now known as prime time. Over time, though, television began to be seen during the day time and late at night, as well on the weekends. As air time increased so did the demand for new material. With the exception of sports television, variety programs became much more important in prime time.

Scheduling Strategies

Block Programming

Block programming occurs when the television network schedules similar programs back-to-back. The concept is to provide similar programming to retain viewership.

Bridging

Bridging is being used when a station tries to prevent the audience from changing channels during a junction point - the main evening breaks where all channels stop programs and shift gear. This is achieved in a number of ways including: having a program already underway and something compelling happening at a junction point, running a program late so that people 'hang around' and miss the start of other programs, or television advertising the next program during the credits of the previous.

Crossprogramming

Crossprogramming involves the interconnection of two shows. This is achieved by extending a storyline over two episodes of two different programs.

Counterprogramming

Counterprogramming is the practice of offering television programs to attract an audience from another television station airing a major event. It is also referred when programmers offer something different from the rival's program as an alternative, to increase the audience size. Is used when a time period is filled with a program whose appeal is different from the opponent program because it is a different genre or appeals to a different demographic.

Dayparting

Dayparting is the practice of dividing the day into several parts, during each of which a different type of radio programming or television programming appropriate for that time is aired. Daytime television programs are most often geared toward a particular demographic, and what the target audience typically engages in at that time.

- Sign-on
- Early morning news
- Early morning
- Late morning
- Daytime television

- Early fringe
- Lunchtime news
- Early afternoon
- Late afternoon
- Early evening
- Evening news
- Prime time
- Late-night news
- Late night television
- Graveyard slot
- Sign-off (closedown)
- Late fringe
- Post late-fringe

Hammocking

Hammocking is a technique used by broadcasters whereby an unpopular program is scheduled between two popular programs in the hope that viewers will watch it. Public television use this as a way of promoting serious but valuable content.

Hotswitching

In hotswitching, the programmers eliminate any sort of commercial break when one program ends and another begins; this immediately hooks the audience into watching the next program without a chance to change the television channel between programs.

Spoiling

Spoiling tactics are used to grab audience share, when broadcasters have similar products going head to head. In such cases broadcasters may jostle in getting a slightly earlier airing date or time, in the hope that once viewers have become committed to a program they will not switch channels.

Stacking

Stacking is a technique used to develop audience flow by grouping together programs with similar appeals to "Sweep" the viewer along from one program to the next.

Stripping

Stripping is running a syndicated television series every day of the week. It is commonly restricted to describing the airing of shows which were weekly in their first run; *The West Wing* could be stripped, but not *Jeopardy!*, as the latter is already a daily show. Shows that are syndicated in this way generally have to have run for several seasons (the rule of thumb is usually 100 episodes) in order to have enough episodes to run without significant repeats.

Tentpoling

In tent pole programming, the programmers bank on a well-known series having so much audience appeal that they can place two unknown series on either side, and it is the strength of the central program that will draw viewers to the two other shows.

Theming

Another strategem is having special theme days, such as for a holiday, or theme weeks such as Discovery Channel's *Shark Week*.

Time Slot

A show's time slot or place in the schedule could be crucial to its success or failure. It may affect casting; for example, ABC replaced Robert Lansing's character on *Twelve O'Clock High* because the show moved from 10 pm to 7:30 pm. Promising new series would often be premiered behind hits to help build an audience. Conversely, failing shows could be consigned to unfavorable times, such as the Friday night death slot.

Scheduling Strategies

A typical scheduling strategy used in Argentinian radio and television is called "pase" (Spanish for a "pass" as in a player passing the ball to another player of the same team). A few minutes before the end of a live broadcast show, followed by another live broadcast show, people from both programmes will share some air time together. This may be used for people from the starting programme to anticipate its contents of the day, or to participate in an ongoing discussion in the previous show, or simply for an entirely independent debate or chat that will not be furthered after the "pase". On the radio, where newscasts are usually broadcast every thirty minutes, often in coincidence with the end of a show, the "pase" may take some minutes before the news, and sometimes some minutes afterwards, too.

Alternatively, if there is no "pase", light jokes or comments can be made in a show involving people of the following show, so that some viewers or listeners might be interested in hearing what the reply will be.

Also, when a station has a new show starting, or if it needs to boost its ratings, part of its cast will be featured in other programmes in the same station, inserted in the dynamics of the programme they are in. For example, they will participate in game shows, be interviewed by the journalists of the station, make cameos in a series, substitute for the usual staff of other shows in their habitual functions, etc. Additionally, hosts of live programmes may mention repeatedly the new show and its time slot, trying to encourage their own viewers to watch it.

Rerun

A rerun or repeat is a rebroadcast of an episode of a radio or television program. There are two types of reruns – those that occur during a hiatus, and those that occur when a program is syndicated. Reruns can also be, as the case with more popular shows, when a show is aired outside of its timeslot (for example, in the afternoon).

In the United Kingdom, the word "repeat" refers only to a single episode; "rerun" or "rerunning" is the preferred term for an entire series/season. A "Repeat" is a single episode of a series that is broadcast outside its original timeslot on the same channel/network. The episode is usually the "repeat" of the scheduled episode that was broadcast in the original timeslot earlier the previous week. It allows viewers who weren't able to watch the show in its timeslot to catch-up before the next episode is broadcast. The term "rerun" can also be used in some respects as a synonym for *reprint*, the equivalent term for print items; this is especially true for print items that are part of ongoing series (such as comic strips; *Peanuts*, for instance, has been in reruns since the retirement and death of creator Charles M. Schulz). In South Africa, reruns of the daily soap opera 7de Laan, and others, are called an Omnibus. The Omnibus is a weekly rerun that is broadcast on a Sunday afternoon on the original channel/network. It only broadcasts the past week's episodes back-to-back.

When used to refer to the rebroadcast of a single episode, Lucille Ball and Desi Arnaz are generally credited as the inventors of the rerun; it was first utilized for the American television series *I Love Lucy* (1951-57) during Ball's pregnancy.

Reruns in the United States

In the United States, most television shows from the late 1940s and early 1950s were performed live, and in many cases they were never recorded. However, television networks in the United States began making kinescope recordings of shows broadcast live from the East Coast. This allowed the show to be broadcast later for the West Coast. These kinescopes, along with pre-filmed shows, and later, videotape, paved the way for extensive reruns of syndicated television series.

During Hiatus

In the United States, currently running shows will rerun older episodes from the same season to fill the time slot with the same program. This is often done for headliner shows because the length of the year (52 weeks) is far more than the length of a pick-up (from six to 13 episodes – usually one per week) or a full season (usually from 22 to 24 episodes or weeks). Shows will tend to start rerunning episodes after the November sweeps period and usually show only reruns from mid-December until mid-January or even February sweeps (where a show will return to airing new episodes in order to increase its ratings, which will determine the cost of a commercial run during that time slot). This winter (or "mid-season") phase is also used to try out new shows that did not make it onto the fall schedule to see how they fare with the public. These series usually run six to 13 episodes. If they do well with the public, they may get a renewal for a half (13 weeks) or full season in the new fall schedule. Major shows that are already a hit with audiences will return from

February sweeps until the end of the season (which sometimes ends before May sweeps) with only limited reruns used.

Television Specials

Often, if a television special such as *Peter Pan* or a network television broadcast of a classic film like *The Wizard of Oz* is especially well-received, it will be rerun from time to time. Before the VCR era, this would be the only opportunity audiences had of seeing a program more than once.

Seasonal programming such as *How the Grinch Stole Christmas, It's A Wonderful Life* or the Charlie Brown television specials are normally re-shown once (or occasionally twice) each year, in the appropriate timeframe.

Syndication

A television program goes into syndication when many episodes of the program are sold as a package for a large sum of money. Generally the buyer is either a cable channel or an owner of local television stations. Often, programs are not particularly profitable until they are sold for syndication. Since local television stations often need to sell more commercial airtime than network affiliates, syndicated shows are usually edited to make room for extra commercials. Often about 100 episodes (four to five seasons' worth) are required for a weekly series to be rerun in daily syndication (at least four times a week). Very popular series running more than four seasons may start daily reruns of the first seasons, while production and airings continue of the current season's episodes; until approximately the early 1980s, shows that aired in syndication while still in production had the reruns aired under an alternate name (or multiple alternate names, as was the case with *Death Valley Days*) to differentiate the reruns from the first-run episodes.

Few people anticipated the long life that a popular television series would eventually see in syndication, so most performers signed contracts that limited residual payments to about six repeats. After that, the actors received nothing and the production company would keep 100% of any income until the copyright expired; many shows did not even have their copyrights renewed and others were systematically destroyed to recycle valuable film, such was the lack of awareness of the potential for revenue from them. This situation went unchanged until the mid-1970s, when contracts for new shows extended residual payments for the performers, regardless of the number of reruns, while tape recycling effectively came to an end (rapid advancements in digital video in the 1990s made preservation far more economical) and the Copyright Act of 1976 extended copyright terms to much longer lengths, eliminating the need for renewal.

Once a series is no longer performing well enough to be sold in syndication, it may still remain in *barter* syndication, in which television stations are offered the program for free in exchange for a requirement to air additional advertisements (without compensation) bundled with the free program during other shows (barter syndication is far more common, if not the norm, in radio, where only the most popular programs charge rights fees). The Program Exchange is the most prominent barter syndicator in United States television, offering mostly older series from numerous network libraries. Barter syndicated series may be seen on smaller, independent stations with small budgets or as short-term filler on larger stations; they tend not to be as widely syndicated as programs syndicated with a rights fee.

Classic Television

With the growing availability of cable and satellite television channels as well as over-the-air digital subchannels, combined with a growing body of available post-syndication programming, a handful of specialty channels have been built solely or primarily to run former network programming which otherwise would no longer be in syndication. Branded as "classic television", these often carry reruns of programming dating back to the monochrome television era and are promoted as nostalgia. The corresponding radio format would be that of an oldies, classic rock, classic hits or adult standards station. Depending on the programs chosen for a classic network, running the format can be very inexpensive, due to many shows beginning to fall into the public domain.

On cable and satellite, channels that devote at least some of their program schedule to post-syndication reruns include Nick at Nite, TV Land, TBS, USA Network, WGN America, TVGN, Hub Network, Game Show Network, Boomerang, Nicktoons, INSP, RFD-TV, and the Hallmark Channel. Equity Media Holdings had been using low-power television stations to carry its own Retro Television Network in various markets; those stations were, as a result of Equity going bankrupt, sold to religious broadcaster Daystar Television Network. Since the early 2010s, the growth of digital subchannel networks has allowed for increasing specialization of these classic networks: in addition to general-interest program networks such as Me-TV, Retro TV and Antenna TV, there exist networks solely for sitcoms (Laff), game shows (Buzzr TV), black-oriented programs (Bounce TV), children's programming (PBJ, qubo), true crime and court programming (Justice Network), and feature films (Movies!, getTV and This TV).

Traditionally, shows most likely to be rerun in this manner are scripted comedies and dramas. Game shows, variety shows, Saturday morning cartoons and, to a lesser extent, newsmagazines and late-night talk shows (often in edited form) have been seen less commonly in reruns. Most variants of reality television have proven to be a comparative failure in reruns, due to the competitive nature of the format and the lack of the element of surprise. Despite this, reruns of sports broadcasts have found a niche, and networks such as MSG Network, ESPN Classic and NFL Network currently have a significant portion of programming time devoted to reruns of live sportscasts.

DVD Retail

With the rise of the DVD video format, box sets featuring season or series runs of television series have become an increasingly important retail item. Some view this development as a rising new idea in the industry of reruns as an increasingly major revenue source in themselves instead of the standard business model as a draw for audiences for advertising. While there were videotape releases of television series before DVD, the format's limited content capacity, large size and reliance on mechanical winding made it impractical as a widespread retail item. Many series which continue to air first-run episodes (such as *Modern Family* and *Grey's Anatomy*) may release DVD sets of the prior season between the end of that season and the beginning of the next.

Some television programs that are released on DVD (particularly those that have been out of production for several years) may not have all of the seasons released, either due to poor overall sales or prohibitive costs for obtaining rights to music used in the program; one such incidence is *Perfect Strangers*, which has seldom been in wide syndication since the late 1990s primarily due to lack of demand, which had only a DVD set of the first and second seasons released due to the expensiveness of relicensing songs used in later seasons of the series that are performed by the show's two lead characters. In some cases, series whose later season releases have been held up for these reasons may have the remaining seasons made available on DVD, often after a distributor that does not hold syndication rights to the program (such as Shout! Factory) secures the rights for future DVD releases.

TV Listings

TV Guide originally used the term *rerun* to designate rebroadcast programs, but abruptly changed to *repeat* in the early 1970s.

Other TV listings services and publications, including local newspapers, would often indicate reruns as "(R)"; since the early 2000s, many listing services now only provide a notation only if an episode is new ("(N)"), with reruns getting no notation.

Repeats in the United Kingdom

In the United Kingdom, most drama and comedy series run for shorter seasons – typically six, seven or thirteen episodes – and are then replaced by others. An exception is soap operas, which are either on all year round (for example, *EastEnders* and *Coronation Street*), or are on for a season similar to the American format.

As in the U.S., fewer new episodes are made during the summer. Until recently it was also common practice for the BBC, ITV and Channel 4 to repeat classic shows from their archives, but this has more or less dried up in favor of newer (and cheaper) formats like reality shows, except on the BBC where older BBC shows, especially sitcoms like *Dad's Army* and *Fawlty Towers*, are frequently repeated.

Syndication did not exist as such in United Kingdom until the arrival of satellite, cable and later, from 1998 on, digital television, although it could be argued that many ITV programs up to the early 1990s, particularly imported programming was syndicated in the sense that each ITV region bought some programs independently of the ITV Network, and in particular many programs out of prime time made by smaller ITV stations were "part-networked" where some regions would show them and others would not. Nowadays there are many channels in the UK (for example, Gold) which repackage and rebroadcast "classic" programming from both sides of the Atlantic. Some of these channels, like their U.S. counterparts, make commercial timing cuts; others get around this by running shows in longer time slots, and critics of timing cuts see no reason why all channels should not do the same.

Early on in the history of British television, agreements with the actors' union Equity and other trade bodies limited the number of times a single program could be broadcast, usually only twice, and these showings were limited to within a set time period such as five years. This was due to the unions' fear that the channels filling their schedules with repeats could put actors and other production staff out of work as fewer new shows would be made. It also had the unintentional side effect of causing many programs to be junked after their repeat rights had expired, as they were considered to be of no further use by the broadcasters. Although these agreements changed during the 1980s and beyond, it is still expensive to repeat archive television series on British terrestrial television, as new contracts have to be drawn up and payments made to the artists concerned. Repeats on multi-channel television are cheaper, as are re-showings of newer programs covered by less strict repeat clauses. However, programs are no longer destroyed, as the historical and cultural reasons for keeping them have now been seen, even if the programs have little or no repeat value.

Simulcast

Simulcast, a portmanteau of simultaneous broadcast, is the broadcasting of programs or events across more than one medium, or more than one service on the same medium, at exactly the same time. For example, Absolute Radio is simulcast on both AM and on satellite radio, and the BBC's Prom concerts were formerly simulcast on both BBC Radio 3 and BBC Television. Another application is the transmission of the original-language soundtrack of movies or TV series over local or Internet radio, with the television broadcast having been dubbed into a local language.

Simulcasting to Provide Stereo Sound for TV Broadcasts

Before stereo TV sound transmission was possible, simulcasting on TV and Radio was a method of effectively transmitting "stereo" sound to music TV broadcasts. In the 1970s WPXI in Pittsburgh broadcast a live Boz Scaggs performance which had the audio simultaneously broadcast on two FM radio stations to create a quadrophonic sound, the first of its kind. The first such transmission was in 1975, when the BBC broadcast a recording of Van Morrison's London Rainbow Concert simultaneously on BBC2 TV and Radio 2.

Similarly, in the 1980s, before Multichannel Television Sound, or home theater was commonplace in American households, broadcasters would air a high fidelity version of a television program's audio portion over FM stereo simultaneous with the television broadcast. PBS stations were the most likely, especially when airing a live concert. It was also a way of allowing MTV and similar music channels to run stereo sound through the cable-TV network. This method required a stereo FM transmitter modulating MTV's stereo soundtrack through the cable-TV network and customers connecting their FM receiver's antenna input to the cable-TV outlet. Then they would tune the FM receiver to the specified frequency that would be published in documentation supplied by the cable-TV provider.

With the introduction of commercial FM stations in Australia in July 1980, commercial TV channels began simulcasting some music based programs with the new commercial FM stations and continued to do so into the early 1990s. These were initially rock based programs, such as late night music video shows and rock concerts, but later included some major rock musicals such as The Rocky Horror Picture Show and The Blues Brothers when they first aired on TV. During the mid 1980s the final Australian concert of several major rock artists such as Dire Straits were simulcast live on a commercial TV and FM station. The ABC also simulcast some programs on ABC Television and ABC FM, including the final concert of Elton John with the Melbourne Symphony Orchestra.

The first ever concert "simulcast" was Frank Zappa's Halloween shows (October 31, 1981), live from NYC's Palladium and shown on MTV with the audio-only portion simulcast over the FM "Starfleet Radio" network. Engineered by Mark Pinske with the UMRK mobile recording truck. A later, notable application for simulcasting in this context was the Live Aid telethon concert that was broadcast around the world on July 13, 1985. Most destinations where this concert was broadcast had the concert simulcast by at least one TV network and at least one of the local FM stations.

Most stereo-capable video recorders made through the 1980s and early 1990s had a "simulcast" recording mode where they recorded video signals from the built-in TV tuner and audio signals from the VCR's audio line-in connectors. This was to allow one to connect a stereo FM tuner that is tuned to the simulcast frequency to the VCR's audio input in order to record the stereo sound of a TV program that would otherwise be recorded in mono. The function was primarily necessary with stereo VCRs that didn't have a stereo TV tuner or were operated in areas where stereo TV broadcasting wasn't in place. This was typically selected through the user setting the input selector to "Simulcast" or "Radio" mode or, in the case of some JVC units, the user setting another "audio input" switch from "TV" or "Tuner" to "Line".

In the mid to late 1990s, video game developer Nintendo utilized simulcasting to provide enhanced orchestral scoring and voice-acting for the first ever "integrated radio-games" – its Satellaview video games. Whereas digital game data was broadcast to the Satellaview unit to provide the basic game and game sounds, Nintendo's partner, satellite radio company St.GIGA, simultaneously broadcast the musical and vocal portion of the game via radio. These two streams were combined at the Satellaview to provide a unified audiotrack analogous to stereo.

Other Uses

The term "simulcast" (describing simultaneous radio/television broadcast) was coined in 1948 by a press agent at WCAU-TV, Philadelphia. NBC and CBS had begun broadcasting a few programs both to their established nationwide radio audience and to the much smaller—though steadily-growing—television audience. NBC's "Voice of Firestone" was an early example. Toscanini's NBC Symphony performance of 15 March 1952 is perhaps a first instance of radio/TV simulcasting of a concert, predating the much-heralded rock concert simulcasts beginning in the 1980s.

Presently, in the United States, simulcast most often refers to the practice of offering the same programming on an FM and AM station owned by the same entity, in order to cut costs. With the advent of solid state AM transmitters and computers, it has become very easy for AM stations to broadcast a different format without additional cost; therefore, simulcast between FM/AM combinations are rarely heard today outside of rural areas, and in urban areas, where often the talk or all-news radio format of an AM station is simulcast on FM, mainly for the convenience of office buildings in urban cores which easily block AM signals.

During apartheid in South Africa, many foreign programmes on SABC television were dubbed in Afrikaans. The original soundtrack, usually in English, but sometimes in German or Dutch was available on the Radio 2000 service. This could be selected using a button labeled simulcast on many televisions manufactured before 1995.

Radio programs have been simulcast on television since the invention thereof however, as of recent, perhaps the most visible example of radio shows on television is *The Howard Stern Show*, which currently airs on Sirius Satellite Radio as well as Howard TV. Another prominent radio show that was simulcast on television is *Imus in the Morning*, which until the simulcast ended in 2015, aired throughout the years on MSNBC, RFD-TV and Fox Business Network, in addition to its radio broadcast distributed by Citadel Media. Multiple sports talk radio shows, including *Mike & Mike*, *The Herd with Colin Cowherd* and *Boomer and Carton* also are carried on television, saving those networks the burden of having to air encores of sporting events or other paid sports programming which may draw lower audiences.

In professional wrestling, a simulcast happened on March 26, 2001 between *WWF Raw is War* and *WCW Monday Nitro* upon WWE's purchase of WCW's assets to merge the storylines of the two wrestling promotions, which was the last episode of *Monday Nitro*.

In another case, popular programs will be aired simultaneously on different services in adjacent countries, such as *The Simpsons*, airing Sunday evenings at 8:00 p.m. (Eastern and Pacific times) on both Fox in the United States and Global in Canada. "Simulcast" is often a colloquial term for the related Canadian practice of simultaneous substitution (simsub).

Simulcasts are also used for the purposes of television ratings, mainly with awards ceremonies such as the MTV Video Music Awards and the Nickelodeon Kids' Choice Awards where the ceremony airs on other sister channels in the same corporate family. These allow a bulk ratings number to be competed which allows for more homes to be calculated in a final rating, along with removing any in-house competition, with each of the networks carrying the same ceremony and same advertising. Another example is a "roadblock strategy", where a family of networks will air the premiere of a new series, music video or other event such as a telethon at the same time to maximize their audiences. During major breaking news events, a simulcast of the Fox News Channel is aired on the Fox broadcast network and Fox Business Network.

"Live Simulcast" is also used throughout South America for a real-time live broadcasting from the USA, where it differs from just a live broadcasting in that, on the former case, the event is being broadcast live while it is happening in real time (e.g. the NFL games), while, the latter, a show may be a live recording but not necessarily being broadcast in real time of when the event took place (e.g. a live concert recording). Yet, local live productions (soccer games, for instance) being broadcast in real time in South America are often just called live, without the use of the word simulcast.

Simulcasting of Sporting Events

In sports, such as American football and baseball, simulcasts are when a single announcer broadcasts play-by-play coverage both over television and radio. The practice was common in the early years of television, but since the 1980s, most teams have used a separate team for television and for radio.

As all NFL television broadcasts are done by the national networks or via cable, there are no regular TV-to-radio football simulcasts. However, NFL rules require that games airing on cable and satellite networks (ESPN, NFL Network) be simulcast on local over-air TV stations in markets serving the two local teams participating in each game.

Similarly, no current National Basketball Association teams use a simulcast. Al McCoy (Phoenix),

Chick Hearn (Los Angeles), Kevin Calabro (Seattle) and Rod Hundley (Utah) were the last NBA team broadcasters to simulcast.

In Major League Baseball, only Vin Scully continues the practice; however, he simulcasts only the first three innings of Los Angeles Dodgers games at Dodger Stadium and other National League Western Division parks.

The practice is more prevalent in the National Hockey League where two teams currently simulcast:

- The Buffalo Sabres with Rick Jeanneret and Rob Ray
- The Dallas Stars with Ralph Strangis and Daryl Reaugh (popularly known as "Ralph and Razor")

Simulcasts via satellite can be a challenge, as there is a significant delay because of the distance - nearly 50,000 miles (80,000 km) round-trip - involved. Anything involving video compression (and to some extent audio data compression) also has an additional significant delay, which is noticeable when watching local TV stations on direct-broadcast satellites. Even though the process is not instantaneous, this is still considered a simulcast because it is not intentionally stored anywhere.

(Multiplexing—also sometimes called "multicasting"—is something of a reversal of this situation, where multiple program streams are combined into a single broadcast. The two terms are sometimes confused.)

In horse racing, a simulcast is a broadcast of a horse race which allows wagering at two or more sites; the simulcast often involves the transmission of wagering information to a central site, so that all bettors may bet in the same betting pool, as well as the broadcast of the race.

The San Francisco Giants simulcast with the Oakland Athletics while playing each other on their respective stations and commercials with a mix of broadcasters from both teams.

On cable television systems, analog-digital simulcasting (ADS) means that analog channels are duplicated as digital subchannels. Digital tuners are programmed to use the digital subchannel instead of the analog. This allows for smaller, cheaper cable boxes by eliminating the analog tuner and some analog circuitry. On DVRs, it eliminates the need for an MPEG encoder to convert the analog signal to digital for recording. The primary advantage is the elimination of interference, and as analog channels are dropped, the ability to put 10 or more SDTV (or two HDTV, or various other combinations) channels in its place. The primary drawback is the common problem of over-compression (quantity over quality) resulting in fuzzy pictures and pixelation.

In universities with multiple campuses, simulcasting may be used for a single teacher to teach class to students in two or more locations at the same time, using videoconferencing equipment.

In many public safety agencies, simulcast refers to the broadcasting of the same transmission on the same frequency from multiple towers either simultaneously, or offset by a fixed number of microseconds. This allows for a larger coverage area without the need for a large number of channels, resulting in increased spectral efficiency. This comes at the cost of overall poorer voice quality, as multiple sources increase multipath interference significantly, resulting in what is called simulcast distortion.

With some of the latest Simulcast control equipment for FM radio networks, the distortion experienced is almost in-audible to the human ear. With the introduction of Line Equalisation Modules and Tone Generation Modules, the phasing advance and retard is so well calculated that the distortion is almost entirely averted.

The Tone Generation Module (or TGM) generates a pilot tone at 3300 Hz which is then sampled by the Line Equalisation Module (or LEM) which each channel on each radio high site has 2 of located back at the main control site. This then determines the phase shift in the signal and adjusts the transmission accordingly such that all the overlap areas in transmission are in phase with each other.

Broadcast Signal Intrusion

Broadcast signal intrusion is the hijacking of broadcast signals of radio, television stations, cable television broadcast feeds or satellite signals. Hijacking incidents have involved local TV and radio stations as well as cable and national networks.

Although television, cable and satellite broadcast signal intrusions tend to receive more media coverage, radio station intrusions are more frequent, as many simply rebroadcast a signal received from another radio station. All that is required is an FM transmitter that can overpower the same frequency as the station being rebroadcast. Other methods that have been used in North America to intrude on legal broadcasts include breaking into the transmitter area and splicing audio directly into the feed.

As a cable television operator connects itself in the signal path between individual stations and the system's subscribers, broadcasters have fallen victim to signal tampering on cable systems on multiple occasions.

Confirmed Events

"Vrillon" on Southern Television

In November 1977, a hoax audio message, purporting to come from outer space, was broadcast during an ITN news bulletin on Southern Television in the UK. The intrusion did not affect the video signal but replaced the programme audio with warnings as to the destiny of the human race and a disaster to affect "your world and the beings on other worlds around you." The IBA confirmed that it was the first time such a hoax transmission had been made. None of the individuals responsible for the intrusion have been identified.

Captain Midnight on HBO

At 12:32 am Eastern Time on April 27, 1986, HBO (Home Box Office) had its satellite signal feed from its operations center on Long Island in Hauppauge, New York interrupted by a man calling himself "Captain Midnight". The interruption occurred during a presentation of *The Falcon and the Snowman*. The intrusion lasted between 4 and 5 minutes and was seen by viewers along the East Coast. The man, who during the interruption also threatened to hijack the signals of Showtime and The Movie Channel, was later caught and identified as John R. MacDougall of Ocala, Florida. He was prosecuted shortly thereafter. Authorities were tipped off by a man from Wisconsin in a phone booth at a rest area off Interstate 75 in Gainesville, Florida. The man filing the report said that he overheard MacDougall bragging about it.

MacDougall's message as seen by HBO viewers.

MacDougall's guilt was confirmed by an FCC investigation that showed he was alone at Central Florida Teleport at the time of the incident and a recording of the jamming video showed that the text was created by a character generator at that location. He was charged with transmitting without a radio license in violation of 47 U.S.C. § 301. MacDougall pleaded guilty and was fined $5,000 and served a year probation. Ambiguity about whether the 47 USC 301 charge was applicable since the transmitter had a license resulted in the passage of 18 U.S.C. § 1367 which made satellite jamming a felony.

MacDougall was able to perform the intrusion while working a second job as a master control operator at a satellite teleport in Florida, where he worked to make ends meet due to declining income from his satellite TV equipment business. He stated that he did it because he was frustrated with HBO's service rates, and that it was hurting his business selling satellite dishes (hence his second job at the teleport). The message, placed over SMPTE color bars, broadcast by MacDougall read:

GOODEVENING HBO
FROM CAPTAIN MIDNIGHT
$12.95/MONTH ?
NO WAY !
[SHOWTIME/MOVIE CHANNEL BEWARE!]

"Religious" Takeover

Playboy's satellite network was intentionally "jammed" with a text-only message on the night of Sunday, September 6, 1987. The message read, "Thus sayeth the Lord thy God. Remember the sabath [sic] and keep it holy. Repent for the kingdom of Heaven is at hand." (The verses were taken from Exodus 20:8 and Matthew 4:17 in the Bible.)

The white text and black background of the message prevented viewers from seeing the movie that Playboy was running at the time called *Three Daughters*. (Playboy's feed was on RCA Satcom 4, transponder 24.)

Thomas Haynie of Virginia Beach, Virginia was convicted of satellite piracy in September 1990 in Norfolk, Virginia federal court in connection with the incident. Haynie was an uplink engineer at the Christian Broadcasting Network, a television ministry in Virginia Beach. He was on duty at the time of the jamming.

According to investigators, it was the religious content of the transmission and the type of equipment used that drew them to CBN. The jamming signal left behind subtle technical clues that were captured on a VHS recording made at Playboy's uplink at the time of the event - like finding "fingerprints" in the video. After investigators were confident that they identified the brand of transmitter and character (text) generator from the video, they concluded that CBN was the culprit, even though they only examined about 5% of those devices used throughout the United States. (Of course, any earth station or uplink truck operating without a license was never examined since they would not have appeared in the FCC's records.) In reality, CBN's equipment was never an exact match. CBN maintained that the FCC's case was entirely circumstantial since there were no witnesses and the signal could not be traced to a point of origin. (The signal could have originated nearly anywhere in the United States.) Furthermore, CBN asserted that it was standard practice to deliberately limit their transmitters to produce only as much power as was necessary to achieve an optimal transmission, but not to generate enough power to completely overwhelm the signal from another uplink. Simply put, there wasn't enough power for Haynie to jam Playboy's signal.

During the investigation, experts on both sides attempted to recreate the incident with CBN's equipment. They were unsuccessful. Further testing of CBN's equipment proved that the transmitter's klystron amplifier tube was incapable of generating sufficient power on transponder 24 to cause the jamming signal. (The tube was subsequently damaged by the prosecution's investigative team, thus destroying a key piece of evidence.)

At the heart of the FCC's case was the extremely broad and questionable assertion that Haynie was a disgruntled religious zealot on a personal crusade against pornography. In spite of evidence and testimony in his favor, Haynie was the first person to be convicted of violating a new federal law against "satellite piracy" (18 USC 1367), passed after the HBO / "Captain Midnight" case in 1986.

After initially being deadlocked, the jury eventually sided with the prosecution and convicted Haynie on two of six counts. (Haynie was also accused of interfering with the American Exxxtacy channel on GTE Spacenet 1 on the same date; however, a recording of the event was of such poor quality that it was unusable and Haynie was acquitted of the associated charges.) Haynie received three years of probation, a $1000 fine, and 150 hours of community service. Considering that he could have received 11 years of jail time and a $260,000 fine, this was anything but a "slam dunk" for the prosecution.

Haynie is still employed at CBN's uplink. He has always maintained his innocence.

Max Headroom Incident

Screen shot of the Max Headroom hijacker.

On November 22, 1987, an unidentified man wearing a Max Headroom mask intercepted the signals of two television stations in Chicago. Independent station WGN-TV (now a CW affiliate), owned by Tribune Company, was hijacked first. Its signal was hijacked during the sports report on its 9:00 pm newscast for about 25 seconds. Then came PBS station WTTW, where the man was seen and heard uttering garbled remarks before dropping his trousers, and was then spanked with a flyswatter before the screen went black. The interception occurred at about 11:00 pm during an episode of *Doctor Who* entitled "Horror of Fang Rock" and lasted almost 90 seconds. To this day, none of the individuals responsible for the intrusion have been identified. This incident got the attention of the CBS Evening News the next day and was talked about nationwide. The HBO incident was also mentioned in the same news report, presented by Frank Currier.

"Telewizja Solidarność" (TV Solidarity)

In September 1985, four astronomers at Poland's University of Toruń, Zygmunt Turło, Leszek Zaleski, Piotr Łukaszewski and Jan Hanasz, used a home computer, a synchronizing circuit, and a transmitter to superimpose messages in support of the labor movement Solidarność (Solidarity) over state-run television broadcasts in Toruń, including an episode of *07 zgłoś się*. The messages read "Enough price increases, lies, and repressions. Solidarity Toruń" and "It is our duty to boycott the election" (referring to the rigged Sejm elections of 1985) with the Solidarity logo. The four men were eventually discovered and were charged with "possession of an unlicensed radio transmitter and publication of materials that could cause public unrest". At their sentencing, the judge noted their prize winning work in the Polish scientific community and gave each of them probation and a fine of the equivalent of US$100 each (or 3,000,000 old złoty, 300 PLN in today's currency).

The Era of Soviet Pirate Broadcasting

Broadcast signal intrusion was a common practice in the USSR during the 1970s and 1980s due to the absence of and high demand for any non-government broadcasting. As early as 1966, there was a report of an incident in the city of Kaluga where an 18 year old broadcast a hoax announcement that nuclear war had broken out with the United States.

In the mid-1970s so many pirates were operating around the city of Arkhangelsk, especially at night, that local people were urged to telephone reports of violators to a special number.

Hijackers using call signs such as "cucumber", "Radio Millimeter", "Green Goat", "Fortune", and others, would overpower the signal on relay stations for wired radio networks in order to transmit their own programming, or transmit into wired radio networks during gaps in normal programming. Even though the incidents appear to have been fairly common according to reports from the BBC, most were not publicly acknowledged for policy reasons. Reports in newspapers typically referred to the hijackers as "radio hooligans broadcasting drivel, rudeness, vulgarity, uncensored expressions, and trashy music." State news organizations also attempted smear campaigns against such pirate broadcasters, claiming that they had interfered with a state frequency used by Aeroflot, "preventing a doctor in an air ambulance from transmitting information about a patient."

2006 Lebanon War

During the 2006 Lebanon War, Israel overloaded the satellite transmission of Hezbollah's Al Manar TV to broadcast anti-Hezbollah propaganda. One spot showed Hezbollah leader Hassan Nasrallah with crosshairs superimposed on his image followed by three gunshots and a voice saying "Your day is coming" and shots of the Israeli Air Force destroying targets in Lebanon.

Other Incidents

Television Signal Intrusions

On January 3, 2007 in Australia, during a broadcast of an episode of the Canadian television series *Mayday* on the Seven Network, an audio loop unexpectedly started playing, clearly saying in an American accent, "Jesus Christ, help us all, Lord." This same voice message continued to repeat itself over and over during the show for a total of six minutes. A spokesman for Seven later denied that the transmission was a prank or a security breach and claimed that the repeated line was actually part of the original broadcast and said, "Jesus Christ one of the Nazarenes," although there is hardly any similarity between the two phrases. Subsequent investigation by independent researchers revealed that the invading transmission was actually from a video taped news broadcast of a civilian truck being ambushed in Iraq. It remains unknown whether or not this was an intentional act of television piracy or a genuine glitch of some sort.

On June 17, 2007, an intrusion incident occurred on Czech Television's Sunday morning programme *Panorama*, which shows panoramic shots of Prague and various locations across the country, especially mountain resorts. One of the cameras, located in Černý Důl in Krkonoše, had been tampered with on-site and its video stream was replaced with the hackers' own, which contained CGI of a small nuclear explosion in the local landscape, ending in white noise. The broadcast looked authentic enough; the only clue for the viewers was the Web address of the artist group Ztohoven, which had already performed several reality hacking incidents before. Czech Television considered legal action against the group, and tourism workers in the area expressed outrage (since the programme serves to promote tourism in the areas shown).

On July 13, 2007, a grainy photo of a man and woman interrupted Washington, D.C. ABC affiliate WJLA's digital or HD signal. The picture was not transmitted over the analog signal, however. The incident was deemed a genuine signal intrusion by various websites but has since been confirmed to be the result of an older HDTV encoder malfunctioning in the early morning hours and going undetected. Station management stated that the image was from an advertisement for *The Oprah Winfrey Show*.

On February 11, 2013, Great Falls, Montana CBS affiliate KRTV had their Emergency Alert System hijacked with an audible message warning viewers that "dead bodies are rising from their graves". Later the same night in Marquette, Michigan, the same type of hijacking and reference to a "zombie invasion" was made over the EAS systems of ABC affiliate WBUP and PBS member station WNMU during primetime programming. Shortly afterwards, PBS affiliate KENW of Portales, New Mexico was struck with a similar hacking incident, repeating similar information regarding zombies; however, this led to the arrest of the hacker of the four television stations.

Cable Network Feed Intrusions

On May 1, 2007, a Comcast headend replaced the Disney Channel's program *Handy Manny* with hard-core pornography for viewers in Lincroft, New Jersey. Comcast stated it was conducting an investigation into the event's cause but did not announce findings to the public.

On February 1, 2009, another Comcast headend, in Tucson, Arizona, replaced NBC affiliate KVOA's signal with graphic footage from the pornographic video *Wild Cherries 5* in portions of Arizona for ten seconds, interrupting Super Bowl XLIII between the Arizona Cardinals and the Pittsburgh Steelers during the fourth quarter. Comcast claimed "Our initial investigation suggests this was an isolated malicious act. We are conducting a thorough investigation to determine who was behind this." KVOA also announced that it will be investigating the incident. On February 4, 2011, former Cox Cable employee Frank Tanori Gonzalez was arrested by the FBI and local police in relation to the case.

On April 20, 2012, three minutes of a gay pornographic film was broadcast during a morning news show on CHCH TV in Hamilton, Ontario, Canada for Shaw cable viewers. The night before, a cable was cut; while it was being fixed on the morning of the incident, the adult programming was spliced into CHCH's feed.

Satellite Feed Intrusions

On September 18, 2012, the Disney Channel once again was interrupted on the Dish Network, replacing 6 minutes of *Lilo & Stitch* with a portion of a hardcore pornographic movie. The incident was reported to Dish Network.

On March 11, 2016, private satellite dish owners in Israel watching *HaAh HaGadol* on Channel 2 had their viewing of adventures of house-bound reality stars interrupted with incitement from Hamas. The disruption lasted a little over three and a half minutes.

References

- Nebeker, Frederik (2009). Dawn of the Electronic Age: Electrical Technologies in the Shaping of the Modern World, 1914 to 1945. John Wiley & Sons. pp. 14–15. ISBN 0470409746.
- Banerjee, Indrajit; Seneviratne, Kalinga, eds. (2006), Public service broadcasting in the age of globalization, Asian Media Information and Communication Centre, ISBN 981-4136-01-8
- Raboy, Marc (1995), Public broadcasting for the 21st century, Acamedia research monographs, 17, Indiana University Press, ISBN 1-86020-006-0

- Linke, Benjamin (2016), Public Financing of Public Service Broadcasting and its Qualification as State Aid, Peter Lang, ISBN 978-3-631-66568-8.
- Price, Monroe Edwin; Raboy, Marc (2003), Public service broadcasting in transition: a documentary reader (PDF), Kluwer Law International, ISBN 90-411-2212-5.
- Vogel, Harold (2011). "Television-programming accounting". Entertainment Industry Economics: A Guide for Financial Analysis (Eighth ed.). Cambridge University Press. p. 213. ISBN 978-1-107-00309-5.
- Meehan, Eileen R. (2005). Why TV is not our fault: television programming, viewers, and who's really in control. Rowman & Littlefield. ISBN 0-7425-2486-8.
- Stilson, Janet (January 26, 2014). "Binge Viewing, Cord Cutting and Streaming Platforms Are All the Rage at NATPE 'The syndicated model is broken'". AdWeek. Mediabistro Holdings. Retrieved June 25, 2015.
- Boardman, Madeline (27 April 2015). "Daytime Emmys 2015 Complete Winners List: Soaps, Talk Shows Win Big". Us Weekly. Retrieved 12 August 2015.
- Bibel, Sara (January 8, 2015). "'Celebrity Name Game' Renewed for Season 2 by Fremantle & Debmar-Mercury". TV By the Numbers (Press release). Zap2it. Retrieved May 24, 2015.
- Uribe, R., Buzeta, C. and Hurtado, D. (2011). "Looking for the audiences: The effect of using partial counterprogramming and a friendlier style of news presentation". INNOVAR 21 (42): 151–159. Retrieved 29 April 2014.
- Plunkett, John (April 3, 2013). "The Voice v Britain's Got Talent: scheduling wars recommence". The Guardian. Retrieved December 5, 2014.
- Brown, Maggie (September 23, 2012). "BBC pilots Tuesday night slot as it takes on ITV in the battle of the costume dramas". The Guardian. Retrieved December 5, 2014.
- "'Lords' and 'Knights' defy Soviet authority, 'pollute' the airwaves - CSMonitor.com". The Christian Science Monitor. January 18, 1984. Retrieved 8 January 2014.
- Lewis, Jerry D. (15–21 May 1965). "The General Died at Dusk / Robert Lansing was fine on 10 o'clock missions, but...". TV Guide. p. 24. Retrieved 21 September 2013.

5

Applications of Electronic Media

Electronic media is used in the fields of digital journalism, broadcast journalism, electronic publishing, mobile content and podcasts. This chapter discusses these applications and the impact of electronic media in revolutionizing the field of communication for the publisher as well as the reader. The topics discussed in the chapter are of great importance to broaden the existing knowledge on electronic media.

Journalism

Journalism is the work and distribution of reports on the interaction of events, facts, ideas, and people that are the "news of the day" and that informs society to at least some degree. The word applies to the occupation (professional or not), the methods of gathering information, and the organizing literary styles. Journalistic media include: print, television, radio, Internet, and, in the past, newsreels.

In some nations, the news media is controlled by government intervention, and is not a fully independent body. In others, the news media is independent from the government but the profit motive is in tension with constitutional protections of freedom of the press. Access to freely available information gathered by independent and competing journalistic enterprises with transparent editorial standards can enable citizens to effectively participate in the political processes.

The role and status of journalism, along with that of the mass media, has undergone profound changes over the last two decades with the advent of digital technology and publication of news on the Internet. This has created a shift in the consumption of print media channels, as people increasingly consume news through e-readers, smartphones, and other electronic devices, challenging news organizations to fully monetize their digital wing, as well as improvise on the context in which they publish news in print. Notably, in the American media landscape, newsrooms have reduced their staff and coverage as traditional media channels, such as television, grapple with declining audiences. For instance, between 2007 and 2012, CNN edited its story packages into nearly half of their original time length.

This compactness in coverage has been linked to broad audience attrition, as a large majority of respondents in recent studies show changing preferences in news consumption. The digital era has also ushered in a new kind of journalism in which ordinary citizens play a greater role in the process of news making, with the rise of citizen journalism being possible through the Internet. Using video camera equipped smartphones, active citizens are now enabled to record footage of news events and upload them onto channels like YouTube, which is often discovered and used by mainstream news media outlets. Meanwhile, easy access to news from a variety of online sources, like blogs and other social media, has resulted in readers being able to pick from a wider choice of official and unofficial sources, instead of only from traditional media organizations.

Forms

Photojournalists photographing President Barack Obama of the USA in November 2013.

There are several different forms of journalism, all with diverse audiences. In modern society, "prestige" journalism is said to serve the role of a "fourth estate", acting as a watchdog on the workings of the government. Other forms of journalism feature in different formats and cater to different audiences.

Photo and broadcast journalists interviewing government official after a building collapse in Dar es Salaam, Tanzania. March 2013.

Some forms include:

- Advocacy journalism – writing to advocate particular viewpoints or influence the opinions of the audience.
- Broadcast journalism – written or spoken journalism for radio or television.
- Citizen journalism -- participatory journalism.
- Data journalism -- journalism done with data.
- Drone journalism – use of drones to capture journalistic footage.
- Gonzo journalism – first championed by Hunter S. Thompson, gonzo journalism is

a "highly personal style of reporting".

- Interactive journalism: a type of online journalism that is visual and allows readers to interact with the story through the web
- Investigative journalism: the use of investigation on a subject matter while uncovering news events.
- Photojournalism: the telling of a story through its images.
- Sensor journalism: the use of sensors to support journalistic inquiry.
- Tabloid journalism – writing that is light-hearted and entertaining.
- Yellow journalism (or sensationalism) – writing which emphasizes exaggerated claims or rumours.

The recent rise of social media has resulted in arguments to reconsider journalism as a process rather than attributing it to particular news products. From this perspective, journalism is participatory, a process distributed among multiple authors and involving journalists as well as the socially mediating public.

History

Johann Carolus's *Relation aller Fürnemmen und gedenckwürdigen Historien*, published in 1605 in Strassburg, is often recognized as the first newspaper. The first successful English daily, the *Daily Courant*, was published from 1702 to 1735. The reform of the *Diário Carioca* newspaper in the 1950s is usually referred to as the birth of modern journalism in Brazil.

Role

Walter Lippmann in 1914

In the 1920s, as modern journalism was just taking form, writer Walter Lippmann and American philosopher John Dewey debated over the role of journalism in a democracy. Their differing philosophies still characterize a debate about the role of journalism in society and the nation-state.

To Lippmann, the journalist fulfilled the role of mediator, or translator, between the general public and policy-making elites. Lippmann reasoned that the public could not assess modern society's growingly complex flurry of facts; therefore, it needed an intermediary to filter its news. Journalists served as this intermediary, recording the information exchanged among elites, distilling it, and passing it on for public consumption. The public would affect the decisions of the elite with its vote; in the meantime, the elite would keep the business of power running. Effectively, Lippmann's philosophy had the public at the bottom of the power chain, inheriting its information from the elite.

Lippmann's elitism had consequences that he came to deplore. An apostle of historicism and scientism, Lippmann did not merely hold that democratic government was a problematic exercise, but regarded all political communities, of whatever stripe, as needing guidance from a transcendent partisanship for accurate information and dispassionate judgment. In "Liberty and the News" (1919) and "Public Opinion" (1921) Lippmann expressed the hope that liberty could be redefined to take account of the scientific and historical perspective and that public opinion could be managed by a system of intelligence in and out of government. Thus the liberty of the journalist was to be dedicated to gathering verifiable facts while commentators like himself would place the news in the broader perspective. Lippmann deplored the influence of powerful newspaper publishers and preferred the judgments of the "patient and fearless men of science". In so doing, he denigrated not only the opinion of the majority but also the opinion of those who had influence or power as well. In a republican form of government, the representatives are chosen by the people and share with them adherence to the fundamental principles and political institutions of the polity. Lippmann's quarrel was with those very principles and institutions, for they are the product of the pre-scientific and pre-historical viewpoint and what for him was a groundless natural-rights political philosophy.

But Lippmann turned against what he called the "collectivism" of the Progressive movement he encouraged with its de-emphasis on the foundations of American politics and government and ultimately wrote a work, "The Public Philosophy" (1955), which came very close to a return to the principles of the American founders.

John Dewey

Dewey, on the other hand, believed not only that the public was capable of understanding the issues created or responded to by the elite, but also that it was in the public forum that decisions should be made after discussion and debate. When issues were thoroughly vetted, then the best ideas would bubble to the surface. Dewey believed journalists should do more than simply pass on information. He believed they should weigh the consequences of the policies being enacted. Over time, his idea has been implemented in various degrees, and is more commonly known as "community journalism".

This concept of community journalism is at the centre of new developments in journalism. In this new paradigm, journalists are able to engage citizens and the experts and elites in the proposition and generation of content. While there is an assumption of equality, Dewey still celebrated expertise. Dewey believed the shared knowledge of many to be far superior to a single individual's knowledge. Experts and scholars are welcome in Dewey's framework, but there is not the hierarchical structure present in Lippmann's understanding of journalism and society. According to Dewey, conversation, debate, and dialogue lie at the heart of a democracy.

While Lippmann's journalistic philosophy might be more acceptable to government leaders, Dewey's approach is a more encompassing description of how many journalists see their role in society, and, in turn, how much of society expects journalists to function. Americans, for example, may criticize some of the excesses committed by journalists, but they tend to expect journalists to serve as watchdogs on government, businesses and actors, enabling people to make informed decisions on the issues of the time.

Elements

Bill Kovach and Tom Rosenstiel propose several guidelines for journalists in their book *The Elements of Journalism*. Because journalism's first loyalty is to the citizenry, journalists are obliged to tell the truth and must serve as an independent monitor of powerful individuals and institutions within society. The essence of journalism is to provide citizens with reliable information through the discipline of verification.

Professional and Ethical Standards

News photographers and reporters waiting behind a cordon in New York City, in May 1994

While various existing codes have some differences, most share common elements including the principles of — truthfulness, accuracy, objectivity, impartiality, fairness and public accountability — as these apply to the acquisition of newsworthy information and its subsequent dissemination to the public.

Some journalistic Codes of Ethics, notably the European ones, also include a concern with discriminatory references in news based on race, religion, sexual orientation, and physical or mental disabilities. The Parliamentary Assembly of the Council of Europe approved in 1993 Resolution 1003 on the Ethics of Journalism which recommends journalists to respect the presumption of innocence, in particular in cases that are still *sub judice.*

In the UK, all newspapers are bound by the Code of Practice of the Press Complaints Commission. This includes points like respecting people's privacy and ensuring accuracy. However, the Media Standards Trust has criticized the PCC, claiming it needs to be radically changed to secure public trust of newspapers.

This is in stark contrast to the media climate prior to the 20th century, where the media market was dominated by smaller newspapers and pamphleteers who usually had an overt and often radical agenda, with no presumption of balance or objectivity.

Because of the pressure on journalists to report news promptly and before their competitors, factual errors occur more frequently than in writing produced and edited under less time pressure. Thus a typical issue of a major daily newspaper may contain several corrections of articles published the previous day. Perhaps the most famous journalistic mistake caused by time pressure was the Dewey Defeats Truman edition of the Chicago Daily Tribune, based on early election returns that failed to anticipate the actual result of the 1948 US presidential election.

Failing to Uphold Standards

Such a code of conduct can, in the real world, be difficult to uphold consistently. Reporting and editing do not occur in a vacuum but always reflect the political context in which journalists, no less than other citizens, operate. Journalists who believe they are being fair or objective may give biased accounts—by reporting selectively, trusting too much to anecdote, or giving a partial explanation of actions. Even in routine reporting, bias can creep into a story through a reporter's choice of facts to summarize, or through failure to check enough sources, hear and report dissenting voices, or seek fresh perspectives. A 2011 study of journalism in online video found that although most news videos adhere to traditional production practices (e.g. editing and audio quality), they tended to use more relaxed standards for content (e.g., use of sources, fairness). Videos using these more relaxed standards received more views.

A news organization's budget inevitably reflects decision-making about what news to cover, for what audience, and in what depth. Those decisions may reflect conscious or unconscious bias. When budgets are cut, editors may sacrifice reporters in distant news bureaus, reduce the number of staff assigned to low-income areas, or wipe entire communities from the publication's zone of interest.

Publishers, owners and other corporate executives, especially advertising sales executives, can try to use their powers over journalists to influence how news is reported and published. Journalists usually rely on top management to create and maintain a "firewall" between the news and other departments in a news organization to prevent undue influence on the news department. One journalism magazine, *Columbia Journalism Review*, has made it a practice to reveal examples of executives who try to influence news coverage, of executives who do not abuse their powers over journalists, and of journalists who resist such pressures.

Because of the need to please many different and sometimes powerful audiences, journalists often make a blanket claim to objectivity, even neutrality, which conveniently coincides with the requirements of the market. Although some analysts point to the inherent difficulty of maintaining objectivity, and others practically deny that it is possible, still others point to the requirements of a free press in a democratic society governed by public opinion and a republican government under a limited constitution; leaving out key information too often happens. According to this latter view, direct or implicit criticism of the government, political parties, corporations, unions, schools and colleges and even churches is both inevitable and desirable, and cannot be done well without clarity regarding fundamental political principles. Hence, objectivity consists both in truthful, accurate reporting and well-reasoned and thoughtful commentary, based upon a firm commitment to a free society's principles of equality, liberty and government by consent.

Legal Status

Journalists at a press conference

Governments have widely varying policies and practices towards journalists, which control what they can research and write, and what press organizations can publish. Some governments guarantee the freedom of the press; while other nations severely restrict what journalists can research or publish.

Journalists in many nations have some privileges that members of the general public do not, including better access to public events, crime scenes and press conferences, and to extended interviews with public officials, celebrities and others in the public eye.

Journalists who elect to cover conflicts, whether wars between nations or insurgencies within nations, often give up any expectation of protection by government, if not giving up their rights to protection by government. Journalists who are captured or detained during a conflict are expected to be treated as civilians and to be released to their national government. Many governments around the world target journalists for intimidation, harassment, and violence because of the nature of their work.

Right to Protect Confidentiality of Sources

Journalists' interaction with sources sometimes involves confidentiality, an extension of freedom of the press giving journalists a legal protection to keep the identity of a confidential informant private even when demanded by police or prosecutors; withholding their sources can land journalists in contempt of court, or in jail.

In the United States, there is no right to protect sources in a federal court. However, federal courts will refuse to force journalists to reveal their sources, unless the information the court seeks is highly relevant to the case and there's no other way to get it. State courts provide varying degrees of such protection. Journalists who refuse to testify even when ordered to can be found in contempt of court and fined or jailed. On the journalistic side of keeping sources confidential, there is also a risk to the journalist's credibility because there can be no actual confirmation of whether the information is valid. As such it is highly discouraged for journalists to have confidential sources.

Broadcast Journalism

Photo and broadcast journalists interviewing government official after a building collapse

Broadcast journalism is the field of news and journals which are "broadcast", that is, published by electrical methods instead of the older methods, such as printed newspapers and posters. Broadcast methods include radio (via air, cable, and Internet), television (via air, cable, and Internet) and the World Wide Web. Such media disperse pictures (static and moving), visual text and sounds.

Scripts for broadcast tend to be written differently from text to be read by the public. For instance, the former is generally less complex and more conversational. Radio and television are designed to be seen and heard sooner and more often than a daily or weekly newspaper.

Description

Broadcast articles can be written as "packages", "readers", "voice-overs" (VO) and "sound on tape" (SOT).

A "sack" is an edited set of video clips for a news story and is common on television. It is typically narrated by a reporter. It is a story with audio, video, graphics and video effects. The news anchor, or presenter, usually reads a "lead-in" (introduction) before the package is aired and may conclude the story with additional information, called a "tag".

A "reader" is an article read without accompanying video or sound. Sometimes an "over the shoulder digital on-screen graphic" is added.

A voice-over, or VO, is a video article narrated by the anchor.

Sound on tape, or SOT, is sound or video usually recorded in the field. It is usually an interview or soundbite.

Radio was the first medium for broadcast journalism. Many of the first radio stations were co-operative community radio ventures not making a profit. Later, radio advertising to pay for programs was pioneered in radio. Later still, television displaced radio and newspapers as the main news sources for most of the public in industrialized countries.

Some of the programming on radio is locally produced and some is broadcast by a radio network, for example, by syndication. The "talent" (professional voices) talk to the audience, including reading the news. People tune in to hear engaging radio personalities, music, and information. In radio news, stories include speech soundbites, the recorded sounds of events themselves, and the anchor or host.

Some radio news might run for just four minutes, but contain 12–15 stories. These new bulletins must balance the desire for a broad overview of current events with the audience's limited capacity to focus on a large number of different stories.

The radio industry has undergone a radical consolidation of ownership, with fewer companies owning the thousands of stations. Large media conglomerates such as Clear Channel Communications own most of the radio stations in the United States. That has resulted in more "niche" formats and the sharing of resources within clusters of stations, de-emphasizing local news and information. There has been concern over whether this concentration serves the public. The opposition says that the range of political views expressed is greatly narrowed and that local concerns are neglected, including local emergencies, for which communication is critical. Automation has resulted in many stations broadcasting for many hours a day with no one on the station premises.

Television

Television (TV) news is considered by many to be the most influential medium for journalism. For most of the American public, local news and national TV newscasts are the primary news sources. Not only the numbers of audience viewers, but the effect on each viewer is considered more persuasive ("The medium is the message"). Television is dominated by attractive visuals (including beauty, action, and shock), with short soundbites and fast "cuts" (changes of camera angle). Television viewing numbers have become fragmented, with the introduction of cable news channels, such as Cable News Network (CNN), Fox News Channel and MSNBC.

A journalist works on location at the Loma Prieta Earthquake in San Francisco's Marina District October 1989.

Local Television

The industry divides local television in North America into media markets. These television markets are defined by viewing area and are ranked by the number of audience viewers. New broadcast journalists generally start in the smaller markets with fewer viewers and move up to larger television stations and television networks after gaining experience. The larger stations usually have more resources and better pay.

United States stations typically broadcast local news three or four times a day: around 4:30–6 am, 11:30 or noon, 5 or 6 pm, and 10 or 11 at night. Most of the nightly local newscasts are 30 minutes, and include sports television and weather. News anchors are shown sitting at a desk in a television studio. The news anchors read teleprompters that contain local interest stories and breaking news. Reporters frequently tell their stories outside the formal television studio in the field, in a remote broadcast setting where Electronic news-gathering (ENG) techniques are used with production trucks. Daytime television or morning shows include more "soft" news and feature pieces, while the evening news emphasizes "hard" news.

News Jobs

News anchors (formerly "anchormen") serve as masters-of-ceremonies and are usually shown facing a professional video camera in a television studio while reading unseen teleprompters. The anchors are often in pairs (co-anchors), who sit side by side and often alternate their reading. Meteorologists stand in front of chroma key backgrounds to describe weather forecasting and show maps, charts and pictures. Reporters research and write the stories and sometimes use video editing to prepare the story for air into a "package". Reporters are usually engaged in electronic field production (EFP) and are accompanied by a videographer at the scenes of the news; the latter holds the camera. the videographer or assistants manage the audio and lighting; they are in charge of setting up live television shots and might edit using a non-linear editing system (NLE). Segment producers choose, research and write stories, as well as deciding the timing and arrangement of the newscast. Associate producer, if any, specialize in other elements of the show such as graphics.

Production Jobs

A newscast director is in charge of television show preparation, including assigning camera and talent (cast) positions on the set, as well as selecting the camera shots and other elements for either recorded or live television video production. The technical director (TD) operates the video switcher, which controls and mixes all the elements of the show. At smaller stations, the Director and Technical Director are the same person.

A graphics operator operates a character generator (CG) that produces the lower third on-screen titles and full-page digital on-screen graphics. The audio technician operates the audio mixing console. The technician is in charge of the microphones, music and audio tape. Often, production assistants operate the teleprompters and professional video cameras and serve as lighting and rigging technicians (grips).

Online Convergence

Convergence is the sharing and cross-promoting of content from a variety of media, all of which, in theory, converge and become one medium. In broadcast news, the internet is a key to convergence. Frequently, broadcast journalists also write text stories for the Web, usually accompanied by the graphics and sound of the original story. Websites offer the audience an interactive form where they can learn more about a story, can be referred to related articles, can offer comments for publication and can print stories at home. Technological convergence also lets newsrooms collaborate with other media, broadcast outlets sometimes have partnerships with their print counterparts.

Digital Journalism

Digital journalism also known as online journalism is a contemporary form of journalism where editorial content is distributed via the Internet as opposed to publishing via print or broadcast. What constitutes 'digital journalism' is debated by scholars. However the primary product of journalism, which is news and features on current affairs, is presented solely or in combination as text, audio, video and some interactive forms, and disseminated through digital media platforms.

Fewer barriers to entry, lowered distribution costs, and diverse computer networking technologies have led to the widespread practice of digital journalism. It has democratized the flow of information that was previously controlled by traditional media including newspapers, magazines, radio, and television.

Some have asserted that greater degree of creativity can be exercised with digital journalism when compared to traditional journalism and traditional media. The digital aspect may be central to the journalistic message and remains, to some extent, within the creative control of the writer, editor, and/or publisher.

Overview

There is no absolute agreement as to what constitutes digital journalism. Mu Lin argues that "Web and mobile platforms demand us to adopt a platform-free mindset for an all-inclusive production approach – create the (digital) contents first, then distribute via appropriate platforms." The repurposing of print content for an online audience is sufficient for some, while others require content created with the digital medium's unique features like hypertextuality. Fondevila Gascón adds multimedia and interactivity to complete the digital journalism essence. For Deuze, online journalism can be functionally differentiated from other kinds of journalism by its technological component which journalists have to consider when creating or displaying content. Digital journalistic work may range from purely editorial content like CNN (produced by professional journalists) online to public-connectivity websites like Slashdot (communication lacking formal barriers of entry). The difference of digital journalism from traditional journalism may be in its reconceptualised role of the reporter in relation to audiences and news organizations. The expectations of society for instant information was important for the evolution of digital journalism . However, it is likely that the exact nature and roles of digital journalism will not be fully known for some time.

History

The first type of digital journalism, called teletext, was invented in the UK in 1970. Teletext is a system allowing viewers to choose which stories they wish to read and see it immediately. The information provided through teletext is brief and instant, similar to the information seen in digital journalism today. The information was broadcast between the frames of a television signal in what was called the Vertical Blanking Interval or VBI.

American journalist, Hunter S. Thompson relied on early digital communication technology beginning by using a fax machine to report from the 1971 US presidential campaign trail as documented in his book Fear and Loathing on the Campaign Trail.

After the invention of teletext was the invention of videotex, of which Prestel was the world's first system, launching commercially in 1979 with various British newspapers such as the Financial Times lining up to deliver newspaper stories online through it. Videotex closed down in 1986 due to failing to meet end-user demand.

American newspaper companies took notice of the new technology and created their own videotex systems, the largest and most ambitious being Viewtron, a service of Knight-Ridder launched in 1981. Others were Keycom in Chicago and Gateway in Los Angeles. All of them had closed by 1986.

Next came computer Bulletin Board Systems. In the late 1980s and early 1990s, several smaller newspapers started online news services using BBS software and telephone modems. The first of these was the Albuquerque Tribune in 1989.

Computer Gaming World in September 1992 broke the news of Electronic Arts' acquisition of Origin Systems on Prodigy, before its next issue went to press. Online news websites began to proliferate in the 1990s. An early adopter was *The News & Observer* in Raleigh, North Carolina which offered online news as Nando. Steve Yelvington wrote on the Poynter Institute website about Nando, owned by The N&O, by saying "Nando evolved into the first serious, professional news site on the World Wide Web". It originated in the early 1990s as "NandO Land". It is believed that a major increase in digital online journalism occurred around this time when the first commercial web browsers, Netscape Navigator (1994), and Internet Explorer (1995). By 1996, most news outlets had an online presence. Although journalistic content was repurposed from original text/video/audio sources without change in substance, it could be consumed in different ways because of its online form through toolbars, topically grouped content, and intertextual links. A twenty-four-hour news cycle and new ways of user-journalist interaction web boards were among the features unique to the digital format. Later, portals such as AOL and Yahoo! and their news aggregators (sites that collect and categorize links from news sources) led to news agencies such as The Associated Press to supplying digitally suited content for aggregation beyond the limit of what client news providers could use in the past.

Also, *Salon*, was founded in 1995. In 2001 the *American Journalism Review* called *Salon* the Internet's "preeminent independent venue for journalism."

In 2008, for the first time, more Americans reported getting their national and international news from the internet, rather than newspapers. Young people aged 18 to 29 now primarily get their news via the Internet, according to a Pew Research Center report. Audiences to news sites continued to grow due to the launch of new news sites, continued investment in news online by conventional news organizations, and the continued growth in internet audiences overall. Sixty-five percent of youth now primarily access the news online.

Mainstream news sites are the most widespread form of online newsmedia production. As of 2000, the vast majority of journalists in the Western world now use the internet regularly in their daily work. In addition to mainstream news sites, digital journalism is found in index and category sites (sites without much original content but many links to existing news sites), meta- and comment sites (sites about newsmedia issues like media watchdogs), and share and discussion sites (sites that facilitate the connection of people, like Slashdot). Blogs are also another digital journalism phenomenon capable of fresh information, ranging from personal sites to those with audiences of hundreds of thousands. Digital journalism is involved in the cloud journalism phenomenon, a constant flow of contents in the Broadband Society.

Prior to 2008, the industry had hoped that publishing news online would prove lucrative enough to fund the costs of conventional newsgathering. In 2008, however, online advertising began to slow down, and little progress was made towards development of new business models. The Pew Project for Excellence in Journalism describes its 2008 report on the State of the News Media, its sixth, as its bleakest ever. Despite the uncertainty, online journalists report expanding newsrooms. They believe advertising is likely to be the best revenue model supporting the production of online news.

Many news organizations based in other media also distribute news online, but the amount they use of the new medium varies. Some news organizations use the Web exclusively or as a secondary outlet for their content. The Online News Association, founded in 1999, is the largest organization representing online journalists, with more than 1,700 members whose principal livelihood involves gathering or producing news for digital presentation.

The Internet challenges traditional news organizations in several ways. Newspapers may lose classified advertising to websites, which are often targeted by interest instead of geography. These organizations are concerned about real and perceived loss of viewers and circulation to the Internet.

Hyperlocal journalism is journalism within a very small community. Hyperlocal journalism, like other types of digital journalism, is very convenient for the reader and offers more information than former types of journalism. It is free or inexpensive.

Impact on Readers

Digital journalism allows for connection and discussion at levels that print does not offer on its own. People can comment on articles and start discussion boards to discuss articles. Before the Internet, spontaneous discussion between readers who had never met was impossible. The process of discussing a news item is a big portion of what makes for digital journalism. People add to the story and connect with other people who want to discuss the topic.

Digital journalism creates an opportunity for niche audiences, allowing people to have more options as to what to view and read.

Digital journalism opens up new ways of storytelling; through the technical components of the new medium, digital journalists can provide a variety of media, such as audio, video, and digital photography.

Digital journalism represents a revolution of how news is consumed by society. Online sources are able to provide quick, efficient, and accurate reporting of breaking news in a matter of seconds, providing society with a synopsis of events as they occur. Throughout the development of the event, journalists are able to feed online sources the information keeping readers up-to-date in mere seconds. The speed in which a story can be posted can affect the accuracy of the reporting in a way that doesn't usually happen in print journalism. Before the emergence of digital journalism the printing process took much more time, allowing for the discovery and correction of errors.

News consumers must become Web literate and use critical thinking to evaluate the credibility of sources. Because it is possible for anyone to write articles and post them on the Internet, the definition of journalism is changing. Because it is becoming increasingly simple for the average person to have an impact in the news world through tools like blogs and even comments on news stories on reputable news websites, it becomes increasingly difficult to sift through the massive amount of information coming in from the digital area of journalism.

There are great advantages with digital journalism and the new blogging evolution that people are becoming accustomed to, but there are disadvantages. For instance, people are used to what they already know and can't always catch up quickly with the new technologies in the 21st century. The goals of print and digital journalism are the same, although different tools are needed to function.

The interaction between the writer and consumer is new, and this can be credited to digital journalism. There are many ways to get personal thoughts on the Web. There are some disadvantages to this, however, the main one being factual information. There is a pressing need for accuracy in digital journalism, and until they find a way to press accuracy, they will still face some criticism.

One major dispute regards the credibility of these online news websites. A digital journalism credibility study performed by the Online News Association compares the online public credibility ratings to actual media respondent credibility ratings. Looking at a variety of online media sources, the study found that overall the public saw online media as more credible than it actually is.

The effects of digital journalism are evident worldwide. This form of journalism has pushed journalists to reform and evolve. Older journalists who are not tech savvy have felt the blunt force of this. In recent months, a number of older journalists have been pushed out and younger journalists brought in because of their lower cost and ability to work in advanced technology settings.

Impact on Publishers

Many newspapers, such as the *New York Times*, have created online sites to remain competitive and have taken advantage of audio, video, and text linking to remain at the top of news consumers' lists.

Newspapers rarely break news stories any more, with most websites reporting on breaking news before the cable news channels. Digital journalism allows for reports to start out vague and generalized, and progress to a better story. Newspapers and TV cable are at a disadvantage because they generally can only put together stories when an ample amount of detail and information are available. Often, newspapers have to wait for the next day, or even two days later if it is a late-breaking story, before being able to publish it. Newspapers lose a lot of ground to their online counterparts, with ad revenue shifting to the Internet, and subscription to the printed paper decreasing. People are now able to find the news they want, when they want, without having to leave their homes or pay to receive the news.

Because of this, many people have viewed digital journalism as the death of journalism. Free advertising on websites such as Craigslist has transformed how people publicize; the Internet has created a faster, cheaper way for people to get news out, thus creating the shift in ad sales from standard newspapers to the Internet. There has been a substantial effect of digital journalism and media on the newspaper industry, with the creation of new business models. It is now possible to contemplate a time in the near future when major towns will no longer have a newspaper and when magazines and network news operations will employ no more than a handful of reporters. Many newspapers and individual print journalists have been forced out of business because of the popularity of digital journalism. The newspapers that have not been willing to be forced out of business have attempted to survive by saving money, laying off staff, shrinking the size of the publications, eliminating editions, as well as partnering with other businesses to share coverage and content. In 2009, one study concluded that most journalists are ready to compete in a digital world and that theses journalists believe the transition from print to digital journalism in their newsroom is moving too slowly. Some highly specialized positions in the publishing industry have become obsolete. The growth in digital journalism and the near collapse of the economy has also led to downsizing for those in the industry.

Students wishing to become journalists now need to be familiar with digital journalism in order to be able to contribute and develop journalism skills. Not only must a journalist analyze their audience and focus on effective communication with them, they have to be quick; news websites are able to update their stories within minutes of the news event.

Critics believe digital journalism has made it easier for individuals who are not qualified journalists to misinform the general public. Many believe that this form of journalism has created a number of sites that do not have credible information. Sites such as PerezHilton.com have been criticized for blurring the lines between journalism and opinionated writing.

Some critics believe that newspapers should not switch to a solely Internet-based format, but instead keep a component of print as well as digital. News publication *The Ann Arbor News*, which ceased print publication in July 2009, is an example of this type of format. The News instead switched to an internet based entity, AnnArbor.com, in order to keep up with the transition from print to web.

Digital journalism allows citizens and readers the opportunity to join in on threaded discussions relating to a news article that has been read by the public. This offers an excellent source for writers and reporters to decide what is important and what should be omitted in the future. These threads can provide useful information to writers of digital journalism so that future articles can be pruned and improved to possibly create a better article the next time around.

Work Outside Traditional Press

The Internet has also given rise to more participation by people who are not normally journalists, such as with Indy Media (Max Perez).

Bloggers write on web logs or blogs. Traditional journalists often do not consider bloggers to automatically be journalists. This has more to do with standards and professional practices than the medium. But, as of 2005, blogging has generally gained at least more attention and has led to some effects on mainstream journalism, such as exposing problems related to a television piece about President George W. Bush's National Guard Service.

Recent legal judgements have determined that bloggers are entitled to the same protections as other journalists subject to the same responsibilities. In the United States, the Electronic Frontier Foundation has been instrumental in advocating for the rights of journalist bloggers.

In Canada, the Supreme Court of Canada ruled that:" A second preliminary question is what the new defence should be called. In arguments before us, the defence was referred to as the responsible journalism test. This has the value of capturing the essence of the defence in succinct style. However, the traditional media are rapidly being complemented by new ways of communicating on matters of public interest, many of them online, which do not involve journalists. These new disseminators of news and information should, absent good reasons for exclusion, be subject to the same laws as established media outlets. I agree with Lord Hoffmann that the new defence is "available to anyone who publishes material of public interest in any medium": Jameel, at para. 54."

Other significant tools of on-line journalism are Internet forums, discussion boards and chats, especially those representing the Internet version of official media. The widespread use of the Internet all over the world created a unique opportunity to create a meeting place for both sides in many conflicts, such as the Israeli-Palestinian conflict and the First and Second Chechen Wars. Often this gives a unique chance to find new, alternative solutions to the conflict, but often the Internet is turned into the battlefield by contradicting parties creating endless "online battles."

Internet radio and podcasts are other growing independent media based on the Internet.

Blogs

With the rise of digital media, there is a move from the traditional journalist to the blogger or amateur journalist. Blogs can be seen as a new genre of journalism because of their "narrative style of news characterized by personalization" that moves away from traditional journalism's approach, changing journalism into a more conversational and decentralized type of news. Blogging has become a large part of the transmitting of news and ideas across cites, states, and countries, and bloggers argue that blogs themselves are now breaking stories. Even online news publications have blogs that are written by their affiliated journalists or other respected writers. Blogging allows readers and journalists to be opinionated about the news and talk about it in an open environment. Blogs allow comments where some news outlets do not, due to the need to constantly monitor what is posted. By allowing comments, the reader can interact with a story instead of just absorbing the words on the screen. According to one 2007 study, 15% of those who read blogs read them for news.

However, many blogs are highly opinionated and have a biasSome are not verified to be true. The Federal Trade Commission (FTC) established guidelines mandating that bloggers disclose any free goods or services they receive from third parties in 2009 in response to a question of the integrity of product and service reviews in the online community.

Citizen Journalism

Digital journalism's lack of a traditional "editor" has given rise to citizen journalism. The early advances that the digital age offered journalism were faster research, easier editing, conveniences, and a faster delivery time for articles. The Internet has broadened the effect that the digital age has on journalism. Because of the popularity of the Internet, most people have access, and can add their forms of journalism to the information network. This allows anyone who wants to share something they deem important that has happened in their community. Individuals who are not professional journalists who present news through their blogs or websites are often referred to as citizen journalists. One does not need a degree to be a citizen journalist. Citizen journalists are able to publish information that may not be reported otherwise, and the public has a greater opportunity to be informed. Some companies use the information that a citizen journalist relays when they themselves can not access certain situations, for example, in countries where freedom of the press is limited. Anyone can record events happening and send it anywhere they wish, or put it on their website. Non-profit and grass roots digital journalism sites may have far fewer resources than their corporate counterparts, yet due to digital media are able to have websites that are technically comparable. Other media outlets can then pick up their story and run with it as they please, thus allowing information to reach wider audiences.

For citizen journalism to be effective and successful, there needs to be citizen editors, their role being to solicit other people to provide accurate information and to mediate interactivity among users. An example can be found in the start up of the South Korean online daily newspaper, OhMy-News, where the founder recruited several hundred volunteer "citizen reporters" to write news articles which were edited and processed by four professional journalists

Legal Issues

One emerging problem with online journalism in the United States is that, in many states, individuals who publish only on the Web do not enjoy the same First Amendment rights as reporters who work for traditional print or broadcast media. As a result, unlike a newspaper, they are much more liable for such things as libel. In California, however, protection of anonymous blog sources was ruled to be the same for both kinds of journalism.

Extra-jurisdictional Enforcement

In Canada there are more ambiguities, as Canadian libel law permits suits to succeed even if no false statements of fact are involved, and even if matters of public controversy are being discussed. In British Columbia, as part of "a spate of lawsuits" against online news sites, according to legal columnist Michael Geist, several cases have put key issues in online journalism up for rulings. Geist mentioned that Green Party of Canada financier Wayne Crookes filed a suit in which he alleged damages for an online news service that republished resignation letters from that party and let users summarize claims they contained. He had demanded access to all the anonymous sources confirming the insider information, which Geist believed would be extremely prejudicial to online journalism. The lawsuit, "Crookes versus openpolitics", attracted attention from the BBC and major newspapers, perhaps because of its humorous name. Crookes had also objected to satire published on the site, including use of the name "gang of Crookes" for his allies. Subsequently, Crookes sued Geist, expanding the circle of liability. Crookes also sued Google, Wikipedia, Yahoo,

PBwiki, domain registrars and Green bloggers who he felt were associated with his political opponents. Crookes' attempt to enforce BC's plaintiff-friendly libel laws on California, Ontario and other jurisdictions led to an immediate backlash in bad publicity but the legal issues remain somewhat unresolved as of November 2009. Crookes lost four times on the grounds that he had not shown anyone in BC had actually read the materials on the minor websites, but this left the major question unresolved: How to deal with commentary deemed fair in one jurisdiction but actionable in another, and how to ensure that universal rights to free speech and reputation are balanced in a way that does not lead to radically different outcomes for two people who might for instance participate in a conversation on the Internet.

Right to Reply or Refactor

International Issues

Non-democratic regimes that do not respect international human rights law present special challenges for online journalism:

- Persons reporting from those regimes or with relatives under those regimes may be intimidated, harassed, tortured or killed and the risk of their exposure generally rises if they become involved in a private dispute and are subjected to civil discovery, or if a plaintiff or police officer or government official pressures an international service provider to disclose their identity.
- If print and broadcast journalists are excluded, unverifiable reports from persons on the spot (as during the Iran election crisis of 2009) may be the only way to relay news at all - each individual incident may be unverifiable though statistically a much more representative sample of events might be gathered this way if enough citizens are participating in gathering the news.
- Court processes that do not explicitly respect the rights of fair comment on public issues, political expression in general, religious freedoms, the right to dissent government decisions or oppose power figures, could be imposed on persons who merely comment on a blog or wiki. If judgments can be enforced at a distance, this may require expensive legal responses or chill on comment while cases move through a remote court, with the proceedings possibly even being heard in a foreign language under rules the commentator never heard of before. If people from relatively free countries engage in conversations with those from oppressive countries, for instance on homosexuality, they may actually contribute to exposing and loss of human rights by their correspondents.

News Collections

The Internet also offers options such as personalized news feeds and aggregators, which compile news from different websites into one site. One of the most popular news aggregators is Google News. Others include Topix.net, and TheFreeLibrary.com.

But, some people see too much personalization as detrimental. For example, some fear that people will have narrower exposure to news, seeking out only those commentators who already agree with them.

As of March 2005, Wikinews rewrites articles from other news organizations. Original reporting remains a challenge on the Internet as the burdens of verification and legal risks (especially from plaintiff-friendly jurisdictions like BC) remain high in the absence of any net-wide approach to defamation.

Electronic Publishing

Electronic publishing (also referred to as e-publishing or digital publishing) includes the digital publication of e-books, digital magazines, and the development of digital libraries and catalogues. Electronic publishing has become common in scientific publishing where it has been argued that peer-reviewed scientific journals are in the process of being replaced by electronic publishing. It is also becoming common to distribute books, magazines, and newspapers to consumers through tablet reading devices, a market that is growing by millions each year, generated by online vendors such as Apple's iTunes bookstore, Amazon's bookstore for Kindle, and books in the Google Play Bookstore. Market research suggests that half of all magazine and newspaper circulation will be via digital delivery by the end of 2015 and that half of all reading in the United States will be done without paper by 2015.

Although distribution via the Internet (also known as online publishing or web publishing when in the form of a website) is nowadays strongly associated with electronic publishing, there are many non network electronic publications such as Encyclopedias on CD and DVD, as well as technical and reference publications relied on by mobile users and others without reliable and high speed access to a network. Electronic publishing is also being used in the field of test-preparation in developed as well as in developing economies for student education (thus partly replacing conventional books) - for it enables content and analytics combined - for the benefit of students. The use of electronic publishing for textbooks may become more prevalent with iBooks from Apple Inc. and Apple's negotiation with the three largest textbook suppliers in the U.S. Electronic publishing is increasingly popular in works of fiction. Electronic publishers are able to respond quickly to changing market demand, because the companies do not have to order printed books and have them delivered. E-publishing is also making a wider range of books available, including books that customers would not find in standard book retailers, due to insufficient demand for a traditional "print run". As well, e-publication is enabling new authors to release books that would be unlikely to be profitable for traditional publishers. While the term "electronic publishing" is primarily used in the 2010s to refer to online and web-based publishers, the term has a history of being used to describe the development of new forms of production, distribution, and user interaction in regards to computer-based production of text and other interactive media.

Process

The electronic publishing process follows some aspects of the traditional paper-based publishing process but differs from traditional publishing in two ways: 1) it does not include using an offset printing press to print the final product and 2) it avoids the distribution of a physical product (e.g., paper books, paper magazines, or paper newspapers). Because the content is electronic, it may be distributed over the Internet and through electronic bookstores, and users can read the material on a range of electronic and digital devices, including desktop computers, laptops, tablet computers, smartphones or e-reader tablets. The consumer may read the published content online a website, in an application on a tablet device, or in a PDF document on a computer. In some cases, the reader may print the content onto paper using a consumer-grade ink-jet or laser printer or via a print on demand system. Some users download digital content to their devices, enabling them to read the content even when their device is not connected to the Internet (e.g., on an airplane flight).

Distributing content electronically as software applications ("apps") has become popular in the 2010s, due to the rapid consumer adoption of smartphones and tablets. At first, native apps for each mobile platform were required to reach all audiences, but in an effort toward universal device compatibility, attention has turned to using HTML5 to create web apps that can run on any browser and function on many devices. The benefit of electronic publishing comes from using three attributes of digital technology: XML tags to define content, style sheets to define the look of content, and metadata (data about data) to describe the content for search engines, thus helping users to find and locate the content (a common example of metadata is the information about a song's songwriter, composer, genre that is electronically encoded along with most CDs and digital audio files; this metadata makes it easier for music lovers to find the songs they are looking for). With the use of tags, style sheets, and metadata, this enables "reflowable" content that adapts to various reading devices (tablet, smartphone, e-reader, etc.) or electronic delivery methods.

Because electronic publishing often requires text mark-up (e.g., Hyper Text Markup Language or some other markup language) to develop online delivery methods, the traditional roles of typesetters and book designers, who created the printing set-ups for paper books, have changed. Designers of digitally published content must have a strong knowledge of mark-up languages, the variety of reading devices and computers available, and the ways in which consumers read, view or access the content. However, in the 2010s, new user friendly design software is becoming available for designers to publish content in this standard without needing to know detailed programming techniques, such as Adobe Systems' Digital Publishing Suite and Apple's iBooks Author. The most common file format is .epub, used in many e-book formats. .epub is a free and open standard available in many publishing programs. Another common format is .folio, which is used by the Adobe Digital Publishing Suite to create content for Apple's iPad tablets and apps.

Academic Publishing

After an article is submitted to an academic journal for consideration, there can be a delay ranging from several months to more than two years before it is published in a journal, rendering journals a less than ideal format for disseminating current research. In some fields such as astronomy and some parts of physics, the role of the journal in disseminating the latest research has largely been replaced by preprint repositories such as arXiv.org. However, scholarly journals still play an important role in quality control and establishing scientific credit. In many instances, the electronic materials uploaded to preprint repositories are still intended for eventual publication in a peer-reviewed journal. There is statistical evidence that electronic publishing provides wider dissemination, because when a journal is available online, a larger number of researchers can access the journal. Even if a professor is working in a university that does not have a certain journal in its library, she may still be able to access the journal online. A number of journals have, while retaining their longstanding peer review process to ensure that the research is done properly, established electronic versions or even moved entirely to electronic publication.

Copyright

In the early 2000s, many of the existing copyright laws were designed around printed books, magazines and newspapers. For example, copyright laws often set out limits on how much of a book can be mechanically reproduced or copied. Electronic publishing raises new questions in relation to copyright, because if an e-book or e-journal is available online, millions of Internet users may be able to view a single electronic copy of the document, without any "copies" being made. Emerging evidence suggests that e-publishing may be more collaborative than traditional paper-based publishing; e-publishing often involves more than one author, and the resulting works are more accessible, since they is published online. At the same time, the availability of published material online opens up more doors for plagiarism or unauthorized use or re-use of the material. Some publishers are trying to address these concerns. For example, in 2011, HarperCollins limited the number of times that one of its e-books can be lent in a public library. Other publishers, such as Penguin, are attempting to incorporate e-book elements into their regular paper publications.

Examples

Electronic versions of traditional media:

- CD-ROM
- E-book
- Electronic journal
- Online magazine
- Online newspaper
- PDF

New media:

- Blog
- Collaborative software
- Digital publication app
- File sharing
- Mobile apps
- Podcast
- Enhanced publication

Business Models

- Digital distribution
- Online advertising
- Open access (publishing)
- Pay-per-view
- Print on demand
- Subscriptions
- Self-publishing
- Non-subsidy publishing

Mobile Content

Mobile content is any type of electronic media which is viewed or used on mobile phones, like ringtones, graphics, discount offers, games, movies, and GPS navigation. As mobile phone use has grown since the mid-1990s, the significance of the devices in everyday life has grown accordingly. Owners of mobile phones can now use their devices to make calendar appointments, send and receive text messages (SMS), listen to music, watch videos, shoot videos, redeem coupons for purchases, view office documents, get driving instructions on a map, and so forth. The use of mobile content has grown accordingly.

Camera phones not only present but produce media, for example photographs with a few million pixels, and can act as pocket video cameras.

Mobile content can also refer to text or multimedia hosted on websites, which may either be standard Internet pages, or else specific mobile pages.

Transmission

Mobile content via SMS is still the main technology for communication used to send mobile consumers messages, especially simple content such as ringtones and wallpapers. Because SMS is the main messaging technology used by young people, it is still the most effective way of reaching this target market. SMS is also ubiquitous, reaching a wider audience than any other technology available in the mobile space (MMS, bluetooth, mobile e-mail or WAP). More important than anything else, SMS is extremely easy to use, which makes adoption increase day by day.

Although SMS is an old technology that may someday be replaced by the likes of Multimedia Messaging Service (MMS) or WAP, SMS frequently gains new powers. One example is the introduction of applications whereby mobile tickets are sent to consumers via SMS, which contains a WAP-push that contains a link where a barcode is placed. This clearly substitutes MMS, which has a limited reach and still suffers from interoperability problems.

It is important to keep enhancing the consumer confidence in using SMS for mobile content applications. This means, if a consumer has ordered a new wallpaper or ringtone, this has to work properly, in a speedy and reliable way. Therefore, it is important to choose the right SMS gateway provider in order to ensure quality-of-service along the whole path of the content SMS until reaching the consumer's mobile.

Modern phones come with Bluetooth and Near field communication. This allows video to be sent from phone to phone over Bluetooth, which has the advantages that there is no data charge.

Content Types

Apps

Mobile application development, also known as mobile apps, has become a significant mobile content market since the release of the first iPhone from Apple in 2007. Prior to the release of Apple's phone product, the market for mobile applications (outside of games) had been quite limited. The bundling of the iPhone with an app store, as well as the iPhone's unique design and user interface, helped bring a large surge in mobile application use. It also enabled additional competition from other players. For example, Google's Android platform for mobile content has further increased the amount of app content available to mobile phone subscribers.

Some examples of mobile apps would be applications to manage travel schedules, buy movie tickets, preview video content, manage RSS news feeds, read digital version of popular newspapers, identify music, look at star constellations, view Wikipedia, and much more. Many televion networks have their own app to promote and present their content. iTyphoon is an example of a mobile application used to provide information about typhoons in the Philippines.

Games

Mobile games are applications that allow people to play a game on a mobile handset. The main categories of mobile games include Puzzle/Strategy, Retro/Arcade, Action/Adventure, Card/Casino, Trivia/Word, Sports/Racing, given in approximate order of their popularity.

Several studies have shown that the majority of mobile games are bought and played by women. Sixty-five percent of mobile game revenue is driven by female wireless subscribers. They are the biggest driver of revenue for the Puzzle/Strategy category; comprising 72 percent of the total share of revenue, while men made up 28 percent. Women dominate revenue generation for all mobile game categories, with the exception of Action/Adventure mobile games, in which men drive 60 percent of the revenue for that category. It's also said that teens are three times as likely as those over twenty to play cell phone games.

Images

Mobile images are used as the wallpaper to a mobile phone, and are also available as screensavers. On some handsets, images can also be set to display when a particular person calls the users. Sites like adg.ms allow users to download free content, however service operators such as Telus Mobility blocks non Telus website downloads.

Music

Mobile music is any audio file that is played on a mobile phone. Mobile music is normally formatted as an AAC (Advanced Audio Coding) file or an MP3, and comes in several different formats. Monophonic ringtones were the earliest form of ringtone, and played one tone at a time. This was improved upon with polyphonic ringtones, which played several tones at the same time so a more convincing melody could be created. The next step was to play clips of actual songs, which were dubbed Realtones. These are preferred by record labels as this evolution of the ringtone has allowed them to gain a cut of lucurative ringtone market. In short Realtones generate royalties for record labels (the master recording owners) as well as publishers (the writers), however, when Monophonic or Polyphonic ringtones are sold only publishing or "mechanical" royalties are incurred as no master recording has been exploited. Some companies promote covertones, which are ringtones that are recorded by cover bands to sound like a famous song. Recently Ringback tones have become available, which are played to the person calling the owner of the ringback tone. Voicetones are ringtones that play someone talking or shouting rather than music, and there are various of ringtones of natural and everyday sounds. Realtones are the most popular form of ringtones. As an example, they captures 76.4% of the US ringtone market in the second quarter of 2006, followed by monophonic and polyphonic ringtones at 12% and ringback tones and 11.5% – but monophonic and polyphonic ringtones are falling in popularity while ringback tones are growing. This trend is common around the globe. A recent innovation is the singtone, whereby "the user's voice is recorded singing to a popular music track and then "tuned-up" automatically to sound good. This can then be downloaded as a ringtone or sent to another user's mobile phone" said the director of Synchro Arts, the developers.

As well as mobile music there are full track downloads, which are an entire song encoded to play on a mobile phone. These can be purchased and bought over the mobile network, but data charges can make this prohibitive. The other way to get a song onto a mobile phone is by "side loading" it, which normally involves downloading the song onto a computer and then transferring it to the mobile phone via Bluetooth, infra-red or cable connections. It is possible to use a full track as a ringtone. In recent years, websites have sprung that allow users to upload audio files and customize them into ringtones using specialized applications, including Myxer, Bongotones, and Zedge.

Mobile music is becoming an integral part of the music industry as a whole. In 2005, the International Federation of Phonographic Industries (IFPI) said it expects mobile music to generate more revenues that online music before the end of that year. In the first half of 2005, the digital music market grew enough to offset the fall in the traditional music market – without including the sale of ringtones, which still makes up the majority of mobile music sales around the globe.

Video

Mobile video comes in several forms including 3GPP, MPEG-4, RTSP, and Flash Lite.

Mobishows and Cellsodes

A Mobishow or a cellsode are terms to describe a broadcast quality programme / series which has been produced, directed, edited and encoded for the mobile phone. Mobishows and Cellsodes can range from short video clips such as betting advice or the latest celebrity gossip, through to half-hour drama serials. Examples include *The Ashes* and *Mr Paparazzi Show* which both were created for mobile viewing.

Streaming

Radio

Mobile streaming radio is an application that streams on-demand audio channels or live radio stations to the mobile phone. In the U.S., mSpot was the first company to develop and commercialize streaming radio which went live in March 2005 on Sprint. Today, all major carriers offer some sort streaming radio service featuring programmed stations based on popular genres and live stations which included both music and talk.

TV

Mobile video also comes in the form of streaming TV over the mobile network, which must be a 2.5G or 3G network. This mimics a television station in that the user cannot elect to see what they wish but must watch whatever is on the channel at the time.

There is also mobile broadcast TV, which operates like a traditional television station and broadcasts the content over a different spectrum. This frees up the mobile network to handle calls and other data usage, and because of the "one-to-many" nature of mobile broadcast TV the video quality is a lot better than that streamed over the mobile networks, which is a "one-to-one" system.

The problem is that broadcast technologies don't have a natural up link, so for users to interact with the TV stream the service has to be closely integrated to the carriers mobile network. The main technologies for broadcast TV are DVB-H, Digital Multimedia Broadcasting (DMB), and MediaFLO.

Live Video

Live video can also be streamed and shared from your cell phone through applications like Qik and InstaLively. The uploaded video can be shared to your friends through emails or social networking sites. Most Live video streaming application works over the cell network or through Wi-Fi. They also require most users to have a dataplan from their cell phone carriers.

International Trends

Since the late 1990s, mobile content has become an increasingly important market worldwide. The South Koreans are the worlds Leaders in Mobile Content and 3-G mobile networks, then the Japanese, followed closely by the Europeans, are heavy users of their mobile phones and have been attaining custom mobile content for their devices for years. In fact, mobile phone use has begun to exceed the use of PCs in some countries. In the United States and Canada, mobile phone use and the accompanying use of mobile content has been slower to gain traction because of political issues and because open networks do not exist in America.

On current trends, mobile phone content will play an increasing role in the lives of millions across the globe in the years ahead, as users will depend on their mobile phones to keep in touch not only with their friends but with world news, sports scores, the latest movies and music, and more.

Mobile content is usually downloaded through WAP sites, but new methods are on the rise. In Italy, 800,000 people are registered users to Passa Parola, an application that allows users to browse a big database for mobile content and directly download it to their handsets. This tool can also be used to recommend content to others, or send content as a gift.

An increasing number of people are also beginning to use applications like Qik to upload and share their videos from their cell phone to the internet. Mobile phone software like Qik allows user to share their videos to their friends through emails, SMS, and even social networking sites like Twitter and Facebook.

Podcast

The *Serial* podcast being played through a podcast player

A podcast is an episodic series of digital media files which a user can set up so that new episodes are automatically downloaded via web syndication to the user's own local computer or portable media player.

The word arose as a portmanteau of "iPod" (a brand of media player) and "broadcast". The files thus distributed are typically audio or video, but may sometimes include other file formats such as PDF or ePub.

The distributor of a podcast maintains a central list of the files on a server as a web feed that can be accessed through the Internet. The listener or viewer uses special client application software on a computer or media player, known as a podcatcher, which accesses this web feed, checks it for updates, and downloads any new files in the series. This process can be automated so that new files are downloaded automatically, which may seem to the user as though new episodes are broadcast or "pushed" to them. Files are stored locally on the user's device, ready for offline use. Podcasting contrasts with webcasting or streaming which do not allow for offline listening, although most podcasts may also be streamed on demand as an alternative to download. Many podcast players (apps as well as dedicated devices) nowadays allow adjusting the playback speed.

Some have labelled podcasting as a converged medium bringing together audio, the web, and portable media players, and also a disruptive technology that has caused some in the radio business to reconsider established practices and preconceptions about audiences, consumption, production, and distribution. Podcasts are usually free of charge to listeners and can often be created for little to no cost, which departs from the traditional model of "gate-kept" media and production tools. It is very much a horizontal media form: producers are consumers, and consumers may become producers and engage in conversations with each other.

Name

"Podcast" is a portmanteau, invented by BBC journalist Ben Hammersley in 2004, of the words "pod" — from iPod, a popular brand of portable media player produced by Apple Inc. — and "broadcast". Despite the etymology, the content can be accessed using any computer or similar device that can play media files. Use of the term "podcast" predated Apple's addition of formal support for podcasting to the iPod, or to Apple's iTunes software.

Other names for podcasting include "net cast", intended as a vendor-neutral term without the loose reference to the Apple iPod. The name is used by shows from the TWiT.tv network. Some sources have suggested the backronym "portable on demand" for "POD", for similar reasons.

History

Many people and groups, including Dawn and Drew of *The Dawn and Drew Show*, Kris and Betsy Smith of *Croncast*, and Dan Klass of *The Bitterest Pill* contributed to the early emergence and popularity of podcasts. Former MTV VJ Adam Curry, in collaboration with Dave Winer, a developer of RSS feeds, is credited with coming up with the idea to automate the delivery and syncing of textual content to portable audio players.

Podcasting, once an obscure method of spreading information, has become a recognized medium for distributing audio content, whether for corporate or personal use. Podcasts are similar to radio programs, but they are audio files. Listeners can play them at their convenience, using devices that have become more common than portable broadcast receivers.

The first application to make this process feasible was iPodderX, developed by August Trometer and Ray Slakinski. By 2007, through the evolution of internet capabilities, along with cheaper hardware and software, audio podcasts were doing what was historically accomplished via radio broadcasts, which since the 1930s had been the sources of radio talk shows and news programs.

In August 2004, Adam Curry launched his show *Daily Source Code*. It was a show focused on chronicling his everyday life, delivering news and discussions about the development of podcasting, as well as promotion for new and emerging podcasts. *Daily Source Code* is believed to be the first podcast produced on a consistent basis. Curry published it in an attempt to gain traction in the development of what would come to be known as podcasting, and as a means of testing the software outside of a lab setting. The name *Daily Source Code* was chosen in the hope that it would attract an audience with an interest in technology.

Daily Source Code started at a grassroots level of production and was initially directed at podcast developers. As its audience became interested in the format, these developers were inspired to create and produce their own projects and, as a result, they improved the code used to create podcasts. As it became known how easy production was, a community of pioneer podcasters quickly appeared. Despite a lack of a commonly accepted identifying name at the time of creation, *Daily Source Code* is commonly believed to be the first podcast to be published online.

In June 2005, Apple released iTunes 4.9 which added formal support for podcasts, thus negating the need to use a separate program in order to download and transfer them to a mobile device. While this made access to podcasts more convenient and widespread, it effectively ended advancement of podcatchers by independent developers. To add to the cooling factor, Apple issued Cease and Desist orders to many podcast application developers and service providers for using the term "iPod" or "Pod" in their products' names.

The logo used by Apple in 2005 to represent podcasting in its iTunes software.

Within a year, many podcasts from public radio networks like the BBC, CBC Radio One, National Public Radio, and Public Radio International placed many of their radio shows on the iTunes platform. In addition, major local radio stations like WNYC in New York City and WHYY-FM radio in Philadelphia, KCRW in Los Angeles placed their programs on their websites and later on the iTunes platform.

Concurrently, CNET, This Week in Tech, and later Bloomberg Radio, the Financial Times, and other for-profit concerns provided podcast content, some using podcasting as their only distribution system.

IP Issues in Trademark and Patent Law

Trademark Applications

In February 10 – 25 March 2005, Shae Spencer Management, LLC of Fairport, New York filed a trademark application to register the term "podcast" for an "online prerecorded radio program over the internet". On September 9, 2005, the United States Patent and Trademark Office rejected the application, citing Wikipedia's podcast entry as describing the history of the term. The company amended their application in March 2006, but the USPTO rejected the amended application as not sufficiently differentiated from the original. In November 2006, the application was marked as abandoned.

As of September 20, 2005, known trademarks that attempted to capitalize on podcast included: ePodcast, GodCast, GuidePod, MyPod, Pod-Casting, Podango, PodCabin, Podcast, Podcast Realty, Podcaster, PodcastPeople, Podgram PodKitchen, PodShop, and Podvertiser.

By February 2007, there had been 24 attempts to register trademarks containing the word "PODCAST" in United States, but only "PODCAST READY" from *Podcast Ready, Inc.* was approved.

Apple Trademark Protections

On September 26, 2004, it was reported that Apple had started to crack down on businesses using the string "POD", in product and company names. Apple sent a cease and desist letter that week to Podcast Ready, Inc., which markets an application known as "myPodder". Lawyers for Apple contended that the term "pod" has been used by the public to refer to Apple's music player so extensively that it falls under Apple's trademark cover. Such activity was speculated to be part of a bigger campaign for Apple to expand the scope of its existing iPod trademark, which included trademarking "IPOD", "IPODCAST", and "POD". On November 16, 2006, the Apple Trademark Department stated that Apple does not object to third party usage of "the generic term" "podcast" to refer to podcasting services, and that Apple does not license the term. However, no statement was made as to whether or not Apple believed they held rights to it.

Personal Audio Lawsuits

Personal Audio, a company referred to as a "patent troll" by the Electronic Frontier Foundation, filed a patent on podcasting in 2009 for a claimed invention in 1996. In February 2013, Personal Audio started suing high-profile podcasters for royalties, including *The Adam Carolla Show* and the *HowStuffWorks* podcast. US Congressman Peter DeFazio's previously proposed "SHIELD Act" intends to curb patent trolls.

In October, 2013, the Electronic Frontier Foundation filed a petition with the US Trademark Office to invalidate the Personal Audio patent.

In August 18, 2014, the Electronic Frontier Foundation announced that Adam Carolla had settled with Personal Audio.

In April 10, 2015, the U.S. Patent and Trademark Office invalidates five provisions of Personal Audio's podcasting patent.

Variants

Enhanced Podcasts

An *enhanced podcast* can display images synchronized with audio. These can contain chapter markers, hyperlinks, and artwork, all of which is synced to a specific program or device. When an enhanced podcast is played within its specific program or device, all the appropriate information should be displayed at the same time and in the same window, making it easier to display materials.

Podcast Novels

A *podcast novel* (also known as a *serialized audiobook* or *podcast audiobook*) is a literary format that combines the concepts of a podcast and an audiobook. Like a traditional novel, a podcast novel is a work of long literary fiction; however, this form of novel is recorded into episodes that are delivered online over a period of time and in the end available as a complete work for download. The episodes may be delivered automatically via RSS, through a web site, blog, or other syndication method. These files are either listened to directly on a user's computer or loaded onto a portable media device to be listened to later.

The types of novels that are podcasted vary from new works from new authors that have never been printed, to well established authors that have been around for years, to classic works of literature that have been in print for over a century. In the same style as an audiobook, podcast novels may be elaborately narrated with separate voice actors for each character and sound effects, similar to a radio play. Other podcast novels have a single narrator reading the text of the story with little or no sound effects.

Podcast novels are distributed over the Internet, commonly on a weblog. Podcast novels are released in episodes on a regular schedule (e.g., once a week) or irregularly as each episode is released when completed, and can either be downloaded manually from a website or blog, or be delivered automatically via RSS or other method of syndication. Ultimately, a serialized podcast novel becomes a completed audiobook.

Some podcast novelists give away a free podcast version of their book as a form of promotion. Some such novelists have even secured publishing contracts to have their novels printed. Podcast novelists have commented that podcasting their novels lets them build audiences even if they cannot get a publisher to buy their books. These audiences then make it easier to secure a printing deal with a publisher at a later date. These podcast novelists also claim the exposure that releasing a free podcast gains them makes up for the fact that they are giving away their work for free.

Video Podcasts

A video podcast on the Crab Nebula by NASA

A *video podcast* (sometimes shortened to "vodcast") includes video clips. Web television series are often distributed as video podcasts.

Dead End Days (2003–2004) is commonly believed to be the first video podcast. That serialized dark comedy about zombies was broadcast from 31 October 2003 through 2004.

Since the spread of the Internet and the use of Internet broadband connection TCP, which helps to identify various applications, a faster connection to the Internet has been created and a wide amount of communication has been created. Video podcasts have become extremely popular online and are often presented as short video clips, usually excerpts of a longer recording. Video clips are being used on pre-established websites, and increasing numbers of websites are being created solely for the purpose of hosting video clips and podcasts. Video podcasts are being streamed on intranets and extranets, and private and public networks, and are taking communication through the Internet to whole new levels.

Most video clips are now submitted and produced by individuals. Video podcasts are also being used for web television, commonly referred to as Web TV, a rapidly growing genre of digital entertainment that uses various forms of new media to deliver to an audience both reruns of shows or series and content created or delivered originally online via broadband and mobile networks, web television shows, or web series. Examples include Amazon, Hulu, and Netflix. Other types of video podcasts used for web television may be short-form, anywhere from 2–9 minutes per episode, typically used for advertising, video blogs, amateur filming, journalism, and convergence with traditional media.

Video podcasting is also helping build businesses, especially in the sales and marketing sectors. Through video podcasts, businesses both large and small can advertise their wares and services in a modern, cost-effective way. In the past, big businesses had better access to expensive studios where sophisticated advertisements were produced, but now even the smallest businesses can create high-quality media with just a camera, editing software, and the Internet.

Oggcast

An oggcast is a podcast recorded and distributed exclusively in the Ogg vorbis audio codec and/or other similarly free codecs. For example, a podcast distributed both in the non-free MP3 format and the free Ogg Vorbis format would not technically meet the definition of an oggcast. In contrast, a podcast distributed in both the Ogg Vorbis and Speex codecs would meet the strict definition of an oggcast. The term oggcast is a combination of the word "ogg" from the term Ogg Vorbis, and the syllable "cast", from "broadcast".

The exact timeline of the term oggcast is uncertain. However, *The Linux Link Tech Show*, one of the longer running Linux podcasts still in production, has a program in the Ogg Vorbis format in its archives from January 7, 2004. Given that a stable release of Ogg Vorbis did not appear until July 19, 2002, it is very likely that the term oggcast was coined sometime between 2002 and 2004.

Oggcasters tend to be broadcasters who prefer not to use audio and video codecs that have patent and/or licensing restrictions, such as the MP3 codec.

Recording and distributing podcasts in the Ogg Vorbis audio format has advantages. Mozilla Firefox and Google Chrome web browsers both support playing Ogg Vorbis files directly in the browser without requiring plugins. Ogg Vorbis may produce better audio quality with a smaller file size than alternative codecs such as AAC or MP3. However, this has not been proven conclusively. Ogg Vorbis is not bound by patents and is considered "free software" in the sense that no corporate entity owns the rights to the format. Some people feel that this is a safer container for their multimedia content for this reason. However, Oggcasters can generally not reach as wide of an audience as more traditional podcasters. This is mainly due to the lack of native Ogg Vorbis support in Microsoft's Internet Explorer and Apple's Safari web browser, and the lack of Ogg Vorbis support in many mobile audio devices.

Oggcast planet maintains a central list of oggcasts.

Uses

Communities use collaborative podcasts to support multiple contributors podcasting through generally simplified processes, and without having to host their own individual feeds. A community podcast can also allow members of the community (related to the podcast topic) to contribute to the podcast in many different ways. This method was first used for a series of podcasts hosted by the Regional Educational Technology Center at Fordham University in 2005. Gronstedt explores how businesses like IBM and EMC use podcast as an employee training and communication channel.

References

- Associated Press (2001). Broadcast News Handbook: A Manual of Techniques and Practices. New York: McGraw-Hill. ISBN 978-0-7136-3882-0.
- Brooks, Brian (2004). Telling the Story: The Convergence of Print, Broadcast and Online Media. Boston: Bedford, St. Martin's Press. ISBN 0-312-40906-0.
- Geoghegan, Michael W.; Klass, Dan (2007). Podcast Solutions: The Complete Guide to Audio and Video Podcasting (2nd ed.). New York: Springer-Verlag New York, Inc. p. 3. ISBN 978-1-59059-905-1.
- Gronstedt, Anders (May 3, 2007). Basics Of Pod Casting (Infoline ASTD) (PDF). ASTD. ASIN 1562864882. ISBN 1-56286-488-2.
- "In Pod We Trust: The man who accidentally invented the word 'podcast'". BBC Radio 4. BBC. 20 November 2015. Retrieved 1 September 2016.
- Herbert, John (2000). Journalism in the Digital Age: Theory and Practice for Broadcast, Print and On-line Media. Taylor & Francis. p. 9. Retrieved February 10, 2014.
- Kawamoto, Kevin (2003). Digital Journalism: Emerging Media and the Changing Horizons of Journalism. Rowman & Littlefield. p. 2. Retrieved February 9, 2014.
- Shiao, Dennis (17 March 2011). "From Association Meetings to Corporate Events, Video is Everywhere". INXPO. Retrieved 8 June 2013.
- "Podcast – Definition and More from the Free Merriam-Webster Dictionary". Merriam-webster.com. 2012-08-31. Retrieved 2014-06-29.
- Grundy, Benjamin (29 September 2007). "How Podcasting Works, History of Podcasting". HowStuffWorks.com; Discovery.com. Retrieved 2012-12-17.
- "Podcasting Community Faces Patent Troll Threat; EFF Wants to Help". Electronic Frontier Foundation. Retrieved 2012-02-28. Personal Audio is claiming that it owns a patent that covers podcasting technology.
- "Marketing your book in the internet age, CreativeChoices interview with John Lenahan". Youtube.com. 2009-08-06. Retrieved 2012-07-20.
- "Free Really Can Make You Money – A Dialogue With Moses Siregar III, David Gaughran interview with Moses Siregar III". Davidgaughran.wordpress.com. 2011-09-05. Retrieved 2012-07-20.
- Shankland, Stephen (2009-05-28). "Google Chrome gets HTML video support | Webware – CNET". News.cnet.com. Retrieved 2012-02-10.

- Escandor Jr., Juan; Barrameda, Shiena (December 22, 2011). "Locally developed app forewarned 'Sendong' fury". Philippine Daily Inquirer. Retrieved January 22, 2012.
- Carlson, David (2003). "History of Online Journalism". In Kawamoto, Kevin. Digital Journalism: Emerging Media and the Changing Horizons of Journalism. Lanham, Md.: Rowman & Littlefield. Retrieved 2011-09-12.
- Rogers, Tony. "What is Hyperlocal Journalism? Sites That Focus On Areas Often Ignored by Larger News Outlets". about.com. Retrieved 2011-09-12.
- Gascón, Fondevila; Francesc, Joan (2010). "El cloud journalism: un nuevo concepto de producción para el periodismo del siglo XXI". Observatorio (OBS*). 4, Núm. 1 (2010): 19–35. ISSN 1646-5954.

Advertising in Electronic Media

6

Electronic media has several advantages over traditional print media and advertising has exploited this. One is bombarded with advertisements on the radio, television, mobile and internet. This chapter focuses on this aspect and educates the reader about the format, design, advantages, disadvantages and techniques used to advertise over electronic media.

Television Advertisement

A television advertisement (variously called a television commercial, commercial or ad in American English, and known in British English as an advert) is a span of television programming produced and paid for by an organization, which conveys a message, typically to market a product or service. Advertising revenue provides a significant portion of the funding for most privately owned television networks. The vast majority of television advertisements today consist of brief advertising spots, ranging in length from a few seconds to several minutes (as well as program-length infomercials). Advertisements of this sort have been used to promote a wide variety of goods, services and ideas since the beginning of television.

Television was still in its experimental phase in 1928, but the medium's potential to sell goods was already predicted.

The effects of television advertising upon the viewing public (and the effects of mass media in general) have been the subject of philosophical discourse by such luminaries as Marshall McLuhan. The viewership of television programming, as measured by companies such as Nielsen Media

Research, is often used as a metric for television advertisement placement, and consequently, for the rates charged to advertisers to air within a given network, television program, or time of day (called a "daypart").

In many countries, including the United States, television campaign advertisements are considered indispensable for a political campaign. In other countries, such as France, political advertising on television is heavily restricted, while some countries, such as Norway, completely ban political advertisements.

The first official, paid television advertisement was broadcast in the United States on July 1, 1941 over New York station WNBT (now WNBC) before a baseball game between the Brooklyn Dodgers and Philadelphia Phillies. The announcement for Bulova watches, for which the company paid anywhere from $4.00 to $9.00 (reports vary), displayed a WNBT test pattern modified to look like a clock with the hands showing the time. The Bulova logo, with the phrase "Bulova Watch Time", was shown in the lower right-hand quadrant of the test pattern while the second hand swept around the dial for one minute. The first TV ad broadcast in the UK was on ITV on 22 September 1955, advertising Gibbs SR toothpaste. The first TV ad broadcast in Asia was on Nippon Television in Tokyo on August 28, 1953, advertising Seikosha (now Seiko), which also displayed a clock with the current time.

General Background

Television advertising involves two main tasks: creating a television advertisement that meets broadcast standards and then, placing the advertisement on television via a targeted air time media buy that reaches the desired customer.

To accomplish the first step means different things in different parts of the world depending on the regulation in place. In the UK for example, clearance must be given by the body Clearcast. Another example is Venezuela where clearance is governed by a body called CNAC. The clearance provides guarantee to the broadcasters that the content of the advertisement meets legal guidelines. Because of this, special extended clearance sometimes applies to food and medical products as well as gambling advertisements.

The second is the process of TV Advertising Delivery and usually incorporates the involvement of a Post-Production House, a Media agency, Advertising Distribution Specialists and the end-goal, the broadcasters.

Characteristics

It is important to choose a television production company and advertising agency with pertinent expertise in these two areas, and it is preferable to choose an agency that both produces advertisements and places air time, because expertise in broadcast quality production and broadcast standards is vital to gaining the advertisement's acceptance by the networks. After the advent of cheap video software and consumer cameras, numerous individuals have offered video production services on the internet. Video production companies that do not regularly place TV advertisements on the air often have their productions rejected by networks for technical or content issues, due to their inexperience with creating broadcast-ready content.

Many television advertisements feature songs or melodies ("jingles") or slogans designed to be striking and memorable, which may remain in the minds of television viewers long after the span of the advertising campaign. Some of these ad jingles or catch-phrases may take on lives of their own, spawning gags that appear in films, television shows, magazines, comics, or literature. These long-lasting advertising elements may be said to have taken a place in the pop culture history of the demographic to whom they appeared. An example is the enduring phrase, "Winston tastes good like a cigarette should", from the eighteen-year advertising campaign for Winston cigarettes from the 1950s to the 1970s. Variations of this dialogue and direct references to it appeared as long as two decades after the advertising campaign expired. Another example is "Where's the Beef?", which grew so popular it was used in the 1984 presidential election by Walter Mondale. Another popular catch-phrase is "I've fallen and I can't get up", which still appears occasionally, over two decades after its first use. Some advertising agency executives have originated more than one enduring slogan, such as Mary Wells Lawrence, who is responsible for such famous slogans as "Raise your hand if you're Sure", "I♥New York" and "Trust the Midas touch."

Advertising agencies often use humor as a tool in their creative marketing campaigns. Many psychological studies have attempted to demonstrate the effects of humor and their relationship to empowering advertising persuasion.

Animation is often used in advertisements. The pictures can vary from hand-drawn traditional animation to computer animation. By using animated characters, an advertisement may have a certain appeal that is difficult to achieve with actors or mere product displays. Animation also protects the advertisement from changes in fashion that would date it. For this reason, an animated advertisement (or a series of such advertisements) can be very long-running, several decades in many instances. Notable examples are the series of advertisements for Kellogg's cereals, starring Snap, Crackle and Pop and also Tony the Tiger. The animation is often combined with real actors. Animated advertisements can achieve lasting popularity. In any popular vote for the most memorable television advertisements in the UK (such as on ITV or Channel 4) the top positions in the list invariably include animations, such as the classic *Smash* and Creature Comforts advertisements.

Other long-running advertising campaigns catch people by surprise, even tricking the viewer, such as the Energizer Bunny advertisement series. It started in the late 1980s as a simple comparison advertisement, where a room full of battery-operated bunnies was seen pounding their drums, all slowing down except one, with the Energizer battery. Years later, a revised version of this seminal advertisement had the Energizer bunny escaping the stage and moving on (according to the announcer, he "keeps going and going and going..."). This was followed by what appeared to be another advertisement: viewers were oblivious to the fact that the following "advertisement" was actually a parody of other well-known advertisements until the Energizer bunny suddenly intrudes on the situation, with the announcer saying "Still going..." (the Energizer Battery Company's way of emphasizing that their battery lasts longer than other leading batteries). This ad campaign lasted for nearly fifteen years. The Energizer Bunny series has itself been imitated by others, via a Coors Light Beer advertisement, in motion pictures, and by current advertisements by GEICO Insurance.

TV Advertisements by Country

United States of America

In the United States, the Nielsen ratings system measures audience viewership of television programs, and provides a way for television broadcasters to determine how popular their television shows are, so that they can decide what rates to charge advertisers for air time.

For each hour in a broadcast day, advertisements take up a fairly consistent proportion of the time. Commercial breaks have become longer. In the 1960s a typical hour-long American show would run for 51 minutes excluding advertisements. Today, a similar program would only be 42 minutes long; a typical 30-minute block of time now includes 22 minutes of programming and eight minutes of advertisements - six minutes for national advertising and two minutes for local.

A television broadcast of the 101-minute film *The Wizard of Oz* (1939) for instance, could, in the early to mid-1960s, take two hours even with advertisements. Today, a telecast of the same film would last between 2½ and three hours, including advertisements. Because of this, it is common practice to edit films to fit within the time allotted.

Over the course of 10 hours, American viewers will be shown approximately three hours of advertisements, twice what they would have seen in the 1960s. If a 1960s show is rerun today, the content may be edited by nine minutes to make room for the extra advertisements.

In the 1950s and 1960s, the average advertisement's length was one minute. As the years passed, the average length shrank to 30 seconds (and sometimes 10 seconds), but more advertisements are now shown during the break.

TV advertisements were coded for identification by broadcasters via an ISCI code. As of March 31, 2014, Ad-ID has been mandated as the standard method of identification for TV advertisements.

A special case of TV advertisements are rare or one time events known as Mega Event Advertising.

Popularity

In the United States, the TV advertisement is generally considered the most effective mass-market advertising format, and this is reflected by the high prices TV networks charge for commercial broadcasting airtime during popular TV events. The annual Super Bowl American football game is known as much for its commercial advertisements as for the game itself, and the average cost of a single 30-second TV spot during this game (seen by 100 million viewers) has reached US$4.5 million (as of February 2013).

It has been suggested that, in general, television executives believe that advertisers covet the 18-49 age demographic and that older viewers are of almost no interest to most advertisers due to their unwillingness to change their buying habits. Products intended for older consumers, such as certain health products and insurance, are advertised regularly on television, generally during programming that appeals to older adults.

The number of viewers within the target demographic is more important to ad revenues than total viewers. According to Advertising Age, during the 2007-08 season, *Grey's Anatomy* was able to charge $419,000 per advertisement, compared to only $248,000 for an advertisement during *CSI*, despite CSI having almost five million more viewers on average. Due to its demographic strength, *Friends* was able to charge almost three times as much for an advertisement as *Murder, She Wrote*, even though the two series had similar total viewer numbers during the seasons they were on the air together. Broadcast networks are concerned by the increasing use of DVRs by young viewers, resulting in aging of the live viewing audience and consequently, lower advertising rates. TV advertisers may also target certain audiences of the population such as certain races, and people of a certain income level or gender. In recent years, shows that tend to target young women tend to be more profitable for advertisements than shows targeted to younger men. This is due to the fact that younger men are watching TV less than their female counterparts.

In the United Kingdom, television advertising is considerably cheaper than in the United States of America. The current record for an advertising slot on British terrestrial television is quoted at being £250,000 for a 30-second slot during the 2010 series of *Britain's Got Talent*. However while British TV advertising is cheaper, this is only to be expected as the United Kingdom has a much lower population (63 million) compared to the US (310 million). So if the £250,000 figure is adjusted into American terms, the population being 5 times bigger and the exchange rate, this figure would become $2 million, much closer to the US Super Bowl figure.

Because a single television advertisement can be broadcast repeatedly over the course of weeks, months, and even years (the Tootsie Roll company has been broadcasting a famous advertisement that asks "How many licks does it take to get to the tootsie center of a Tootsie Pop?" for over three decades), television advertisement production studios often spend very large sums of money in the production of a single thirty-second television spot. This significant expenditure has resulted in a number of high-quality advertisements with high production values, the latest in special effects technology, the most popular personalities, and the best music. A number of television advertisements are so elaborately produced that they can be considered miniature thirty-second movies; indeed, many film directors have directed television advertisements both as a way to gain exposure and to earn a paycheck. One of film director Ridley Scott's most famous cinematic moments was a television advertisement he directed for the Apple Macintosh computer, that was broadcast in 1984. Although this advertisement was broadcast only once (aside from occasional appearances in television advertisement compilation specials and one 1 a.m. airing in Idaho a month before the Super Bowl so that the advertisement could be submitted for the 1983 Clio Awards), it has become famous and well-known, to the point where it is considered a classic television moment.

Despite the popularity of some advertisements, many consider them to be an annoyance for a number of reasons. The main reason may be that the sound volume of advertisements tends to be higher (and in some cases much higher) than that of regular programming. The United States Congress passed a bill on September 30, 2010, called the CALM Act, to reduce the sound volume of advertisements, and loudness rules set by the FCC are effective as of December 13, 2012. In the UK, the Broadcast Committee of Advertising Practice has a similar regulation. The increasing number of advertisements, as well as overplaying of the same advertisement, are secondary annoyance factors. A third might be that television is currently the main medium to advertise, prompting advertising campaigns by everyone from cell-phone companies, political campaigns, fast food restaurants, to local businesses, and small businesses, prompting longer commercial breaks. Another reason is that advertisements often cut into certain parts in the regular programming that are either climaxes of the plot or a major turning point in the show, which many people find exciting or entertaining to watch.

From a cognitive standpoint, the core reason people find advertisements annoying is that the advertisement's offer is not of interest at that moment, or the presentation is unclear. A typical viewer has seen enough advertisements to anticipate that most advertisements will be bothersome, prompting the viewer to be selective in their viewing. Conversely, if an advertisement concerns a topic important to the viewer (such as an advertisement for debt relief shown to a viewer who has received a late notice in the mail), or has entertainment value beyond the basic message (such as the classic humorous spots for Wendy's "Where's the beef?" campaign), then viewers tend to stay with the advertisement, perhaps even looking forward to viewing it again.

Restrictions

Television commercial in 1948

Beginning on January 2, 1971, advertisements featuring cigarettes were banned from American TV. Advertisements for alcohol products are allowed, but the consumption of any alcohol product is not allowed in a television advertisement (for example, an actor in a beer commercial cannot be shown actually drinking the beer). The Federal Trade Commission and the Federal Communications Commission have laid out regulations for television advertising, outlining restrictions on certain products, content, and claims, in addition to mandating minimum technical standards. Additional content standards are set by individual television broadcast entities to accommodate local laws, community standards, and their particular audience demographic; these broadcast outlets examine each incoming advertisement through a process known as "clearance."

Advertisements as Programming

Since the 1960s, media critics have claimed that the boundaries between "programming" and "advertisements" have been eroded to the point where the line is blurred nearly as much as it was during the beginnings of the medium, when almost all individual television shows were sponsored entirely by a single corporation (the model which was carried over from old-time network radio). For much of the 1970s, 1980s, and 1990s, the FCC imposed a rule requiring networks that broadcast programming on Saturday morning and Sunday nights at 7 PM/6 PM Central to air bumpers ("We'll return after these messages...", "...now back to our programming" and variations thereof) to help younger audiences distinguish programs from advertisements. The only programs that were exempt from this rule were news shows and information shows relating to news (such as *60 Minutes*). Conditions on children's programming have relaxed to an extent since the period of the 1970s and 1980s.

Europe

In many European countries television advertisements appear in longer, but less frequent advertising breaks. For example, instead of 3 minutes every 8 minutes, there might be around 6 minutes every half-hour. European Union legislation limits the time taken by commercial breaks to 12 minutes per hour (20%), with a minimum segment length of 20 or 30 minutes, depending on the program content. Live imported telecasts with shorter segments, such as U.S. sporting events, replace the original ads with promotional material. Advertising broadcast time can vary within the EU and other countries and between networks depending on local policy. Unlike in the United States, in Europe the advertising agency name may appear at the beginning or at the end of the advert.

United Kingdom

In the UK, the BBC is funded by a licence fee and does not screen adverts. On the commercial channels, the amount of airtime allowed by the UK broadcasting regulator Ofcom for advertising is an overall average of 7 minutes per hour, with limits of 12 minutes for any particular clock hour (8 minutes per hour between 6pm and 11pm). With 42-minute American exports to Britain, such as *Lost*, being given a one-hour slot, nearly one third of the slot is taken up by adverts or trailers for other programs. Live imported television programs such as *WWE Raw* show promotional material in place of U.S. advert breaks. Infomercials (known as "admags") were originally a feature of the regional commercial ITV stations from launch in 1955 but were banned in 1963.

The first advert to be shown in the UK was an advert for S.R. Toothpaste on September 22, 1955 on the ITV network (its first day).

On 1 July 2000, TV broadcasters began requiring commercials to be delivered to them in widescreen, an event referred to as C-Day in industry promotion of the change.

In 2008, Ofcom announced a review of television advertising and teleshopping regulations, with a view to possibly changing their code, *Rules on the Amount and Distribution of Advertising* (RADA), which regulates the duration, frequency and restriction of adverts on television.

Television advertising specialist, Nick Illston, of advertising-buying agency Pace Media, states that ITV's £250,000 asking price for a 30-second slot during the 2010 series of *Britain's Got Talent* is currently the most expensive advertising slot on television.

Germany

As in Britain, in Germany, public television stations own a major share of the market. Their programming is funded by a licence fee as well as advertisements on specific hours of the day (20 minutes per day; not after 8 p.m.), except on Sundays and holidays. Private stations are allowed to show up to 12 minutes of advertisements per hour with a minimum of 20 minutes of programming in between interruptions.

Greece

In Greece, where most television stations and channels do not remove their logo before a commercial break, it is mandatory for public and commercial TV stations to separate the advertisements from the rest of the program with each station's ident. Subliminal advertising, tobacco and medication advertisements are not allowed. Most programs (series, documentaries, etc.) can have one commercial break every 20 minutes. Advertisements must not exceed the 15% of the broadcasting day or 20% if there are direct advertisements for buying or renting products (like informercials). A commercial break must not exceed the length of 4 minutes with the exceptions of movies/telefilms with the length exceeding the 45 minutes, which can have one commercial break for 9 minutes and of the movies/telefilms that exceed the length of 110 minutes which can have more than one commercial break every 45 minutes for 9 minutes.

France

The Conseil supérieur de l'audiovisuel allows up to 9 minutes of advertising per hour on average in a day. Private channels can only broadcast one commercial break if the show is less than an hour and two commercial breaks if the show is more than an hour. For public channels, advertising is forbidden after 8 p.m.

Ireland

Filming a movie ad

In Ireland, the Broadcasting Authority of Ireland determines the number of adverts on commercial and community TV stations, while the Minister for Communications, Energy and Natural Resources is responsible for the advertising limits of Public Service Broadcasters on TV and Radio, and commercial radio advertising is governed through statue.

Up to 2010 commercial broadcasters where permitted a maximum of 15% advertising time vs. overall broadcast time. In 2010 TV3 asked for an increase in advertising minutes due to the high level of competition from advertising opt-outs coming into Ireland via satellite and cable pay providers, and also the number of live ITV programmes carried by TV3 (such as The X Factor). The BAI agreed to increasing the number of ads on commercial TV broadcasts to bring Irish commercial TV channels into line with UK commercial TV channels. The Independent Broadcasters of Ireland also lobbied for a similar increase, but the regulator was unable to increase their advertising minutes as these are written into Statue, meaning that any change is the responsibility of the members of the national parliament (The House of the Oireachtas).

Commercial Broadcasters may not exceed 18% of their broadcast day on advertising and they may only have a maximum of 12 minutes of adverts per hour, except for children's programming where advertising may not exceed 10 minutes per hour (commercial TV broadcasters include TV3 and Setanta). Irish community TV channels rarely show advertising; however they are permitted to show 6 minutes of advertising per hour.

RTÉ TV and Radio carry a maximum of 9 minutes per hour of advertising, but only an average of 6 minutes of advertising per hour, and the same rules apply to TG4: this amounts to 10% of total broadcast time. On television, RTÉ cannot carry advertising on RTÉ News Now and RTÉjr; they do not carry advertising on programming for the under 6s, RnaG or their Digital Radio services.

Finland

In Finland, there are two mainstream non-commercial channels run by the state owned broadcasting company Yle, that run advertisements only on very infrequent occasions, such as important sport events. The three main commercial channels MTV3, Sub (a subsidiary of MTV3), and Nelonen, all run their advertisements during breaks approximately every 15 minutes. Since digital TV has been introduced, the number of TV channels has grown, with Yle and the main broadcasters all adding new channels (including some subscription channels). Analogue broadcasts ceased in August 2007 and the nation's TV services are now exclusively digital. A typical break lasts about 4 minutes. The length of *individual* advertisements can vary from a few seconds (7, 10 and 15 are common), and they are currently rarely over one minute in length. Many advertisements of supranational companies are commonly dubbed from English language advertisements. Although Swedish is the other official language of Finland, the advertisements do not feature Swedish subtitles nor are any Swedish language advertisements shown with the infrequent exception of some political advertisements at the time of elections.

Russia

The Russian advertising break consists of 2 parts: federal adverts and regional adverts. The duration for each is 4 minutes and 15 minutes per hour respectively. Like Greece, Russia is one of the few countries where the television channel's logo is retained during the commercial break. In Russia, tobacco advertising is prohibited, and in 2013, this was followed suit by alcohol and medication advertising.

Denmark

The Danish DR-channels are funded by a television licence, so they do not show any advertisements. The other Danish television network, TV2 shows advertisements only in blocks between the programs. These can take from 2 minutes to 10 minutes depending on the time to the next show. In Denmark, commercial breaks are strictly prohibited and advertising targeted to children is restricted. Channels like Kanal 5 and TV3 are allowed to interrupt programs, as these channels are being broadcast via satellite from the United Kingdom.

Norway

Television advertising is regulated by the Broadcasting Act. Public service broadcaster NRK is licence-funded, and cannot carry advertising, however sponsorship of programs (typically sport events) is allowed. Commercial channels may interrupt television programmes with ads, but certain conditions apply. The maximum amount of advertising is 15% per hour (09 min). The nation's second-most watched channel - TV2 - is funded by television advertisements. Norwegian-language programs broadcast from abroad - such as TV3 Norway broadcasting from London in the Norwegian language - must comply to British regulations. Political TV advertisements are strictly prohibited, as are all advertisements for tobacco and alcohol. TV advertisements are monitored by the Norwegian Media Authority (Medietilsynet).

Croatia

In Croatia, the commercial broadcasters may show up to 12 minutes of advertising per hour, while the public broadcaster HRT, being partially funded by a license fee, shows less advertising and usually does not interrupt series and films. Tobacco advertising is banned but alcohol and non-prescription medications can be advertised on TV. Political advertisements are allowed only in the short official campaign period before an election and only in separate blocks from other advertisements.

Poland

In Poland, national public broadcaster Telewizja Polska is partially funded by a license fee, advertising breaks can be broadcast only between shows, but for live sports broadasting which contain breaks under the rules of the playing and other events contains breaks can also be interrupted, while almost all commercial broadcasters (except e.g. Canal+) can interrupt shows for advertising broadcasting. Tobacco and alcohol drink advertising are banned respectively by Health Protection Against the Consequences of Tobacco Use and Tobacco Articles Act of November 9, 1995 and Education in Sobriety and Counteracting Alcoholism Act of October 26, 1982 but beer advertising is restricted and can be shown between 8:00 PM and 6:00 AM as well as unhealthy food cannot be broadcast since January 1, 2015 on children's channel such as MiniMini+, TVP ABC and Nickelodeon and on children's show by agreement.

Asia-Pacific

India

In India, the Code for Self-Regulation of Advertising in India, established by the Advertising Standards Council of India (ASCI), is applicable on Television Commercials (TVCs). If consumers see an advertisement which they consider misleading or offensive, they can write to ASCI.

Malaysia

In Malaysia, a government department monitors television advertisements rather than a private institution. Almost all television stations and channels in the country, whether government-owned, private or pay television channels, broadcast advertisements, with television channel logos removed before the start of the commercial break.

There are usually two commercial breaks in a half-hour program and three commercial breaks in an hour-long program, with the exception of news programs. Terrestrial television can only broadcast advertisements during the program that was currently aired except before announcing the breaking of fast in the month of Ramadan, although advertisements may be included during CJ Wow Shop on Media Prima television channels such as NTV7 and TV9 as of June 2016.

Because Malaysian television advertisements were controlled by the government department, the Malaysian TV advertising permit number appears during the first few seconds of any advertisement. The abbreviation at the first part of the permit number which refers to the ministry which handles the TV advertising permit number always changes sometime after the Malaysian government made a new portfolio of ministers and name changes to some ministries after the general election. The usage of advertising code began in 1995, with KP/YYYY/XXXX of which KP means Ministry of Information, YYYY is the year the advertisement produced and the XXXX is the number of the advert permit. This was replaced by KPKK (Ministry of Information, Communications and Culture) and KKMM (Ministry of Communications and Multimedia of Malaysia) in 2009 and 2013, respectively. The advertising code used on channels other than RTM may or may not be used. Other advertising permits includes the KKLIU (Ministry of Health, the Medicine Advertising Authority) for medicinal advertisements, which was used before 1995 and the JIRP (Pesticide Advertising Department) for pesticide advertisements. All pesticide advertisements must show the word "*INI ADALAH IKLAN RACUN PEROSAK*" (This is a pesticide advertisement) and JIRP advertising code at the beginning of the advertisement and the words "*BACALAH LABEL KELUARAN SEBELUM MENGGUNAKANNYA*" (Read the label before use) at the end of the advertisement. It was also used in advertisements on newspapers and magazines.

Astro is also known to delay incoming satellite feeds by two to five minutes (some channels delay by one hour) from the actual time of the start and the end of program (for example a program aired at 1:30 pm will be started at 1:33 pm) with broadcasting advertisements during in-between programs for its purpose of commercial replacement, as government laws forbid advertisements produced from overseas, with exceptions such as advertisement for movies (since release dates may include more than one country).

Liquor advertisements shown after 10:00 pm during non-Malay programs have been banned in the country since 1995, while cigarette advertisements have been banned from showing cigarette packaging since 1995, and have been banned entirely since 2003. Fast-food advertisements during children's programs were also banned in 2007. There are also restrictions on Malaysian television advertisements such as advertisements for 18-rated films, women's care products and unhealthy foods not being allowed to be broadcast during children's programs, and lottery advertising, which is prohibited during Malay programs. Lingerie advertisements are prohibited in Malaysian television, but allowed in non-Malay magazines published in Malaysia.

The Philippines

Philippines advertising industry is self-regulated. The Philippine advertising industry adopted an

Advertising Code of Ethics essentially to promote efficiency in processing applications and resolution cases and avoid costly litigation in regular courts. The earliest Advertising Code of Ethics dates back to the Philippine Board of Advertising (PBA) established in 1974. In 1989, the PBA was renamed Advertising Board of the Philippines (ADBOARD) and was mandated by the Implementing Rules and Regulations of R.A. 7394 or The Consumer Protection Act to ensure that all advertising materials conform to its Code of Ethics. The ADBOARD Advertising Content & Regulations Committee (ACRC) had been the main implementing arm of advertising self-regulation in the Philippines until March 31, 2008 when the Ad Standards Council (ASC) took over this function. The Association of Broadcasters of the Philippines, a self-regulatory organization representing most television and radio broadcasters in the country, limit advertising to 18 minutes per hour, a move taken to help "promote public interest."

In this case some TV broadcasters have to remove TV logos (like ABS-CBN, TV5 & GMA) before starting a commercial break. Also, all commercial content and advertisements will start (right after for most TV shows will break) before or after TV program announcements/etc. Some TV commercials to be aired requires ASC clearance if the content has trivial facts and testimonies.

Since television was introduced to the Philippines in 1953, they used imported TV advertisements until 1960, the same way in which they used billboard advertisements during the American period. In the 1960, P&G paved their way to start the first local TV advertisement.

In 1966, when the Philippine TV turned from black-and-white to color, Colgate-Palmolive was the first to advertise in color.

Some ads have the following notice indicated at the during or at the end of the commercial depending on type of products.

- "Drink Moderately/Responsibly." for alcoholic beverages.
- "Member: BancNet, PDIC Maximum deposit insurance for each depositor P 500,000, PANA Truth in Advertising" for banks
- "GOVERNMENT WARNING: CIGARETTE SMOKING IS DANGEROUS TO YOUR HEALTH." (formerly "Government Warning: Cigarette smoking is hazardous to health" from January 1, 1983 to November 30, 1990) for tobacco products (radio and television advertising of cigarette and hand-rolling tobaccos is banned since July, 2008 in accordance of Section 22 of Republic Act 9,211 or the Tobacco Regulation Act of 2003)
- "MAHALAGANG PAALALA: ANG (name of product) AY HINDI GAMOT AT HINDI DAPAT GAMITING PANGGAMOT SA ANUMANG URI NG SAKIT." (formerly "NO APPROVED THERAPEUTIC CLAIMS") for food, dietary, fiber and herbal supplements.
- "If symptoms persist, consult your doctor." for over-the-counter medicines.
- "This is a paid advertisement." for political ads.
- "Political advertisement paid for and by (name), (address)." also for political ads.
- "The use of milk supplements must only be upon advice of a health professional". (formerly "Breastmilk is the best for babies up to 2 years of age and beyond.") for infant formulas. Used in conjunction with the DOH-IAC permit number.

- "This product is not intended for babies 6 months of age and below". for food/infant device that are targeted to 6 months old and above.
- "Per (DTI-FTEB SPD/DOH-FDA CDRR/CFRR/DA-NMIS) Permit Number (number of permit), Series of (year)" for promos.

Australia

Similar to the European Union, advertising on Australian commercial television is restricted to a certain amount in a 24-hour period, but there are no restrictions on how much advertising may appear in any particular hour. Australian television has one of the highest proportions of advertising content in the world. In primetime there can be 18 minutes or more of advertising (inc. news updates) per hour. Furthermore, product advertisements with informational content are labeled "public service announcements" and not included in the time restrictions; similarly with "this program brought to you by..." announcements, and station identifications. Consequently, there may be less than 40 minutes of actual program time per hour. Foreign, older television programs and movies are noticeably cut; comedy shows often return from an ad break into laughter, for instance. Australia is also one of the few countries in the world where advertisements may appear prior to, and over the top of, the closing credits of a program. There are some restrictions on television advertising in Australia, such as a complete ban on advertising for cigarettes, as well as advertising during programs intended for young children. The ABC, the nation's public broadcaster, broadcasts no external advertisements, but between programs will broadcast promotions for its own programs and merchandise, although is restricted to approximately five minutes per hour. SBS had similar restrictions on advertising until 2005 when it began airing external advertisements between programs, and later, during programs, as per the commercial networks, although still limited to 5 minutes per hour with the first advertisement being half-hourly news updates. Back in 1978, major commercial networks has the longest ever ad break run with the record 53 minutes of advertising.

New Zealand

All major New Zealand television channels, both state-owned and private, screen advertisements. Adverts on average taking up 15 minutes of each hour; there are usually two advert breaks in a half-hour program, and four advert breaks in an hour-long program.

Television advertising began in New Zealand in April 1961, ten months after the first official television broadcast, and was initially allowed only on Tuesdays, Thursdays and Saturdays. Today, television adverts are only prohibited on Good Friday, Easter Sunday and Christmas Day The Advertising Standards Authority is responsible for advertisement compliance, and deals with advertisement complaints (except for election advertising, in which the Broadcasting Standards Authority is responsible).

South Korea

In South Korea, a television advertisement is called a CF, an abbreviation for "commercial film".

Under the current rules, terrestrial channels cannot take in-program commercial breaks. So, the advertisements are usually put between the intro and the start of a program, and between the end credits and the end of the program. Terrestrial channels often divide some longer-length films like *The Ten Commandments* into parts and consider each part as an individual program. Terrestrial channels can take commercial breaks during breaks in action during sporting events.

Pay-television channels can take in-program commercial breaks, although some pay TV channels schedule advertisements in the same way that terrestrial channels do. Regulations for advertisements on terrestrial channels are more strict than those for pay channels. Non-South Korean channels are not subject to these regulations. Tobacco advertisements are prohibited.

Other *SEA* countries and the rest of the Asia-Pacific region

Depending on its style, system and other factors while in the process of broadcasting commercial breaks and TV advertisements, some officials (like government, associations, institutions, etc.) regulate and monitor the assets and broadcasting TV advertisements based on international standards.

Latin America

Argentina

Since September 2010, all Argentine television channels (including cable channels operated from the country itself), are required to separate advertising from the rest of the programming using bumpers with the text "Espacio publicitario" ("Advertising space"). As of June 2016, this usage is no longer compulsory.

Commercial advertising is limited to 12 minutes per hour. In-program advertising is allowed, but counted toward the 12-minute quota. That means that if a 60-minute show has 2 minutes of in-program advertising, the commercial breaks have to be limited to 10 minutes for that specific hour, otherwise the station might face a fine.

From March 2014, advertisements shown in Argentinian television began to use the word "Aviso publicitario de producción nacional" (*Advertisement of domestic production*) and "Aviso publicitario de producción extranjera" (*Advertisement of foreign production*) for local and foreign advertisements, respectively. Also, advertisements for alcoholic products had the warning "Beber con moderación. Prohibida su venta a menores de 18 años" (*Drink in moderation. Do not sell to persons under 18*). The warning is not used in alcohol advertising before 1997.

Use of Popular Music

Prior to the 1970s, music in television advertisements was generally limited to jingles and incidental music; on some occasions lyrics to a popular song would be changed to create a theme song or a jingle for a particular product. An example of this is found on the recent popular Gocompare.com advert that utilises "Over There", the 1917 song popular with United States soldiers in both world wars and written by George M. Cohan during World War I. In 1971 the converse occurred when a song written for a Coca-Cola advertisement was re-recorded as the pop single "I'd Like to Teach the World to Sing" by the New Seekers, and became a hit. Additionally songwriter Paul Williams composed a piece for a Crocker Bank commercial which he lengthened and The Carpenters recorded as "We've Only Just Begun". Some pop and rock songs were re-recorded by cover bands for use in advertisements, but the cost of licensing original recordings for this purpose remained prohibitive in certain countries (including the U.S.) until the late 1980s.

The use of previously recorded popular songs in American television advertisements began in earnest in 1985 when Burger King used the original recording of Aretha Franklin's song "Freeway of Love" in a television advertisement for the restaurant. This also occurred in 1987 when Nike used the original recording of The Beatles' song "Revolution" in an advertisement for athletic shoes. Since then, many classic popular songs have been used in similar fashion. Songs can be used to concretely illustrate a point about the product being sold (such as Bob Seger's "Like a Rock" used for Chevy trucks), but more often are simply used to associate the good feelings listeners had for the song to the product on display. In some cases the original meaning of the song can be totally irrelevant or even completely opposite to the implication of the use in advertising; for example Iggy Pop's "Lust for Life", a song about heroin addiction, has been used to advertise Royal Caribbean International, a cruise ship line. Music-licensing agreements with major artists, especially those that had not previously allowed their recordings to be used for this purpose, such as Microsoft's use of "Start Me Up" by the Rolling Stones and Apple Inc.'s use of U2's "Vertigo" became a source of publicity in themselves.

In early instances, songs were often used over the objections of the original artists, who had lost control of their music publishing, the music of Beatles being perhaps the most well-known case; more recently artists have actively solicited use of their music in advertisements and songs have gained popularity and sales after being used in advertisements. A famous case is Levi's company, which has used several one hit wonders in their advertisements (songs such as "Inside", "Spaceman", and "Flat Beat"). In 2010, research conducted by PRS for Music revealed that "Light & Day" by The Polyphonic Spree is the most performed song in UK TV advertising.

Sometimes a controversial reaction has followed the use of some particular song on an advertisement. Often the trouble has been that people do not like the idea of using songs that promote values important for them in advertisements. For example, Sly and the Family Stone's anti-racism song, "Everyday People", was used in a car advertisement, which angered some people.[who?]

Generic scores for advertisements often feature clarinets, saxophones, or various strings (such as the acoustic/electric guitars and violins) as the primary instruments.

In the late 1990s and early 2000s, electronica music was increasingly used as background scores for television advertisements, initially for automobiles, and later for other technological and business products such as computers and financial services. Television advertising has become a popular outlet for new artists to gain an audience for their work, with some advertisements displaying artist and song information onscreen at the beginning or end.

Future of TV Advertisements

Digital Television Recorders and Advertisement Skipping

After the video cassette recorder (VCR) became popular in the 1980s, the television industry began studying the impact of users fast forwarding through commercials. Advertising agencies fought the trend by making them more entertaining. The introduction of digital video recorders (also known as digital television recorders or DTRs), such as TiVo, and services like Sky+, Dish Network and

Astro MAX, which allow the recording of television programs into a hard drive, also enabled viewers to fast-forward or automatically skip through advertisements of recorded programs.

Though advertisements for cigarettes are banned in many countries, such advertising could still be seen in the sponsorship of events such as auto racing.

There is speculation that television advertisements are threatened by digital video recorders as viewers choose not to watch them. However evidence from the UK shows that this is so far not the case. At the end of 2008 22 percent of UK households had a DTR. The majority of these households had Sky+ and data from these homes (collected via the SkyView panel of more than 33,000) shows that, once a household gets a DTR, they watch 17 percent more television. 82 percent of their viewing is to normal, linear, broadcast TV without fast-forwarding the ads. In the 18 percent of TV viewing that is time-shifted (i.e. not watched as live broadcast), viewers still watch 30 percent of the ads at normal speed. Overall, the extra viewing encouraged by owning a DTR results in viewers watching 2 percent more ads at normal speed than they did before the DTR was installed.

The SkyView evidence is reinforced by studies on actual DTR behaviour by the Broadcasters' Audience Research Board (BARB) and the London Business School.

Product Placement

Other forms of TV advertising include product placement advertising in the TV shows themselves. For example, *Extreme Makeover: Home Edition* advertises Sears, Kenmore, and Home Depot by specifically using products from these companies, and some sports events like the Sprint Cup of NASCAR are named after sponsors, and race cars are frequently covered in advertisements. Many major sporting venues in North America are named for commercial companies, dating back as far as Wrigley Field. Television programs delivered through new mediums such as streaming online video also bring different opportunities to the traditional methods of generating revenue from television advertising.

Overlay Advertisements

Another type of advertisement shown increasingly, mostly for advertising TV shows on the same channel, is an ad overlay at the bottom of the TV screen, which blocks out some of the picture. "Banners", or "Logo Bugs", as they are called, are referred to by media companies as Secondary Events (2E). This is done in much the same way as a severe weather warning is done, only these happen more frequently. They may sometimes take up only 5 to 10 percent of the screen, but in the extreme, they can take up as much as 25 percent of the viewing area. Subtitles that are part of the program content can be completely obscured by banners. Some even make noise or move across the screen. One example is the 2E ads for Three Moons Over Milford, which was broadcast in the months before the TV show's premiere. A video taking up approximately 25 percent of the bottom-left portion of the screen would show a comet impacting into the moon with an accompanying explosion, during another television program.

Google in the TV Advertising Business

In 2007, Google Executive Chairman Eric Schmidt, then CEO, announced plans to enter the television advertising business, despite its lack of any internal video production or network placement capability or expertise. Initial industry speculation was that they would use a business strategy similar to their scheme for radio ad sales, primarily the acquisition of operations system support provider dMarc. Google announced the shutdown of its TV ads product in 2012. According to the Google corporate blog, Google abandoned its radio advertising sales feature and its radio advertisers in February 2009. Google sold its radio buying assets in August 2009. Google's radio effort was widely considered a failure, as was their foray into print advertising sales. These two traditional media schemes were among a number of Google's media and technology experiments, which have yielded mixed results.

Interactive Advertisements

Online video directories are an emerging form of interactive advertising, which help in recalling and responding to advertising produced primarily for television. These directories also have the potential to offer other value-added services, such as response sheets and click-to-call, which enhance the scope of the interaction with the brand.

Shorter Commercial Breaks

During the 2008-09 TV season, Fox experimented with a new strategy, which the network dubbed "Remote-Free TV". Episodes of Fringe and Dollhouse contained approximately ten minutes of advertisements, four to six minutes fewer than other hour-long programs. Fox stated that shorter commercial breaks keep viewers more engaged and improve brand recall for advertisers, as well as reducing channel surfing and fast-forwarding past the advertisements. However, the strategy was not as successful as the network had hoped and it is unclear whether it will be continued in the future.

Competition from Internet

The growing popularity of the internet continues to draw audiences away from advertisers populating just the television platform.

Radio Advertisement

In the US, commercial radio stations make most of their revenue selling “airtime” to advertisers. Of total media expenditures, radio accounts for 6.9%. Radio advertisements or “spots” are available when a business or service provides valuable consideration, usually cash, in exchange for the station airing their spot or mentioning them on air. The United States Federal Communications Commission (FCC), established under the Communications Act of 1934, regulates commercial broadcasting, and the laws regarding radio advertisements remain relatively unchanged from the original Radio Act of 1927, enacted to deal with increasing problems of signal interference as more and more stations sprung up around the country.

History

The first radio broadcasts aired in the early 1900s. However, it wasn’t until 1919 that radio stations began to broadcast continuously, similar to the modern practice. In the United States, on November 2, 1920, KDKA Pittsburgh became the first radio station to receive a commercial license from the government. KDKA engineer Frank Conrad may have been the first to broadcast a radio advertisement on his own experimental station in 1919 when he thanked a Pittsburgh music store on the air for supplying him with phonograph records, although that was for trade, not cash.

Most early radio stations were put on the air by the manufacturers of radio equipment, such as Westinghouse and General Electric or by department stores which sold radios, such as Gimbel’s, Bamberger’s and Wanamaker’s. After all, if they wanted customers to buy radio sets, there had to be stations for them to listen to. At first, they weren’t concerned with the stations making a profit, if sets were being sold. But as more stations began operating, station owners were increasingly faced with the issue of how to maintain their stations financially, because operating a radio station was a significant expense.

In February 1922, AT&T announced it would begin selling “toll broadcasting” to advertisers, in which businesses would underwrite or finance a broadcast, in exchange for being mentioned on the radio. WEAF of New York (now WFAN) is credited with airing the first paid radio commercial, on August 28, 1922, for the Queensboro Corporation, advertising a new apartment complex in Jackson Heights, Queens, near the just-completed #7 subway line. However, it appears other radio stations may actually have sold advertising before WEAF. As early as May 1920, an amateur radio broadcaster leased out his “station” in exchange for $35 per week for twice-weekly broadcasts. And, in Seattle, Washington, Remick’s Music Store purchased a large ad in the local newspaper advertising radio station KFC, in exchange for sponsorship of a weekly program, in March 1922. Additionally, on April 4, 1922, a car dealer, Alvin T. Fuller, purchased time on WGI of Medford Hillside, Massachusetts, in exchanges for mentions.

During radio’s Golden Age, advertisers sponsored entire programs, usually with some sort of message like “We thank our sponsors for making this program possible”, airing at the beginning or end of a program. While radio had the obvious limitation of being restricted to sound, as the industry developed, large stations began to experiment with different formats. Advertising had become a hot commodity and there was money to be made. The advertising director at Shell Oil Co., urged radio broadcasters to deal directly with relevant advertisers, and sell tie-in commercial spots for established radio programs. Like newspaper ads at the time, Sanders figured that advertisers and radio would both benefit from selling ad spots to get the attention of listeners. Radio was an already prominent medium, but Sanders referred to his initiative as radio 'growing up' in terms of its business aspects and how it dealt with advertising. The "visual" portion of the broadcast was supplied by the listener's boundless imagination. Comedian and voice actor Stan Freberg demonstrated this point on his radio show in 1957, using sound effects to dramatize the towing of a 10-ton maraschino cherry by the Royal Canadian Air Force, who dropped it onto a 700-ft. mountain of whipped cream floating in hot-chocolate filled Lake Michigan, to the cheering of 25,000 extras. The bit was later used by the USA's Radio Advertising Bureau to promote radio commercials.

The radio industry has changed significantly since that first broadcast in 1920, and radio is big business today. Although other media and new technologies now place more demands on consumer's time, 95% of people still listen to the radio every week. Internet radio listening is also growing, with 13 percent of the U.S. population listening via this method. Although consumers have more choices today, 92 percent of listeners stay tuned in when commercials break into their programming.

Formats

There is a broad range of choices for type and length of radio commercials. With changes in the radio industry and better production technologies, the mode of commercial presentation has changed, and commercial advertisements can take on a wide range of forms. The two primary types of radio ads are "live reads" and produced spots.

Cousin to the ad-libbed commercial, live read refers to when a DJ reads an advertiser's spot on the air, delivered from a script, fact sheet or personal knowledge. It can also refer to when the DJ "endorses" the advertiser's goods or services. The Radio Advertising Bureau defines an endorsement as: "A commercial in which the...program personality personally recommends an advertiser's product or service, often done live during the program."

Produced spots appear to be more common. A spot is 'produced' if the radio station or an advertising agency record it for the client. Produced commercial formats include: straight read with sound effects or background music, dialogue, monologue (where the voice talent portrays a character, as opposed to an announcer), jingles, and combinations of these. Studies show that the quality of the commercials is as important to listeners, generally, as the number of ads they hear.

Research and Rates

Radio stations today generally run their advertising in clusters or sets, scattered throughout the broadcasting hour. Studies show that the first or second commercial to air during a commercial break has higher recall than those airing later in the set.

Nielsen Audio is one of the primary providers of ratings data in the United States. Most radio stations and advertising agencies subscribe to this paid service, because ratings are key in the broadcast industry. Ad agencies generally purchase radio based on a target demographic. For example, their client may want to reach men between 18 and 49 years old. The ratings enable advertisers to select a specific segment of the listening audience and purchase airtime accordingly. Ratings are also referred to as "numbers" in the business.

The numbers can show who is listening to a particular station, the most popular times of day for listeners in that group, and the percentage of the total listening audience that can be reached with a particular schedule of advertisements. The numbers also show exactly how many people are listening at each hour of the day. This allows an advertiser to select the strongest stations in the market with specificity and tells them what times of day will be the best times to run their ads.

Besides the basic numbers, most radio stations have access to other data, such as Scarborough, that details more about the listening audience than just what age group they fall into. For example, some data will provide the types of activities listeners participate in, their ethnicity, what type of employment they do, their income levels, what kinds of cars they drive, and even whether or not they have been to a particular entertainment venue, for example.

Radio stations sell their airtime according to dayparts. Typically, a station's daypart lineup will look something like the following: 6am-10am, 10am-3pm, 3pm-7pm, and 7pm- midnight. The spots running after midnight, from 12am-6am, are referred to as "overnights". Though this schedule of dayparts can vary from station to station, most stations run similar daypart lineups and sell their advertisements accordingly. Drive times, or morning and evenings when people are commuting, are usually the most popular times of day and also when each station has the most listeners. The "rates", or what the station charges the advertiser, will reflect that.

Rates can also be affected by the time of year an advertiser runs. January is almost always a very slow time of year, and many stations run specials on their rates during that month. This is not the case in warm weather markets like Florida, where "snow birds" migrate and increase population. In this situation rates are usually at their highest as the population swells. The cost of radio advertising also varies on how well the parties negotiate. During busier times of the year, stations can actually sell out of ads entirely, because, unlike the print media, radio stations only have a limited number of commercial units available per hour. During the dot-com boom, some stations ran as much as twenty minutes of ads per hour. While commercial levels are nowhere near as high today, with the average station running approximately nine minutes of ads per hour, peak periods can and do sell out.

Thus, advertising rates will vary depending on time of year, time of day, how well the station does in the particular demographic an advertiser is trying to reach, how well a station does compared to other stations, and demand on station inventory. The busier the time of year for the station, the more an advertiser can expect to spend. And, the higher ranked a station is in the market, according to the ratings data, the more an advertiser can expect to get charged to run on that station.

Advertising rates can vary depending on the length of spot the advertisers elects to run. Although sixty second spots are the most common, stations also sell airtime in thirty, fifteen, ten and two second intervals. Thirty-second ads have always been popular in television advertising, but radio stations just adopted this format recently. Clear Channel kicked off the "Less is More" initiative in 2004, utilizing thirty-second commercials in markets across the US. Though studies show that fewer commercials cause better recall rates, research indicates that traditional sixty-second spots may be the better option, with higher brand and message recall than the newer thirty-second ads.

Stations also run ten-second spots, or "billboards". Typically, this type of spot runs adjacent to some station feature, such as the traffic report, stating, "This traffic brought to you by...", and is usually limited to about thirty words. Fifteen-second spots are generally reserved for station promotional announcements, although some stations sell them.

In addition to traditional radio advertising, some stations are selling airtime during their streaming broadcasts. In the past, the radio station stream included only the commercials that were also running on air. CBS announced it would begin airing 'live reads' in its streaming radio broadcasts, sold and voiced separately from the stations' regular spots, noting the efficacy of live endorsements.

Efficacy

More than eight out of ten Americans feel listening to commercials in exchange for free radio is a "fair deal". Furthermore, broadcast radio advertising often offers the advantage of being localized and inexpensive in comparison with other mediums such as television. Thus, radio advertising can be an effective, low-cost medium through which a business can reach their target consumer. Studies show that radio ads create emotional reactions in listeners. In turn, consumers perceive the ads as more relevant to them personally, which can lead to increased market awareness and sales for businesses running ad schedules. Twenty-five percent of listeners say they're more interested in a product or business when they hear about it on their preferred station.

Local DJs create a personal relationship with their listening audience, and that audience is more likely to trust what they say and respond to their message. Live endorsements are growing in popularity, as advertisers seek new means to reach consumers and cut through the surrounding clutter. Studies show that live reads have recall and response rates higher than the typical recorded spot. Perhaps because of the relationship listeners develop with their favorite station, twenty-six percent of listeners are more interested in a product or business when a DJ endorses it. As more advertisers turn to live endorsements, heavy demand is placed on DJs to announce them. And, as the number of available DJs shrinks, those that are left are often inundated with requests to do endorsements.

Regulatory Considerations

The Communications Act of 1934 established the FCC, which, in turn, regulates the broadcasting media. In response to the payola scandal of the 1950s, the FCC established guidelines, known as "sponsorship disclosure" rules. These rules apply to both pay-for-play and advertisements alike, recognizing that consumers deserve to know "by whom they are being persuaded". Of particular note to advertisers is Section 317 of the Act, also known as the "Announcement of Payment For Broadcast" provision. It states, in general terms, that where money, services or other valuables are directly or indirectly paid to a radio station, in exchange for mentions on the air, the station must disclose that fact. For example, if a cell phone provider gives a free cell phone to a station's DJ, who then talks about that cell-phone provider on the air, perhaps mentioning what great service he has, the announcer must disclose that it is an advertisement.

Usually, listeners are able to discern radio advertisements from entertainment content. The Communications Act does include an "obviousness" exception: where it is obvious that something is a commercial, the announcement-for-payment provision does not apply. However, where anything of value has changed hands in exchange for mentions on the air, a station has a duty to disclose it.

The Federal Trade Commission is also responsible for broadcast industry regulation, in terms of false or misleading advertising practices. In October 2009, the FTC published guidelines regarding endorsements, requiring clear disclosure of the connection between an advertiser and an endorser. Under the FTC guidelines, an endorser is responsible for disclosing any "material connections" they have with a seller/business.

In recent years, there has been a trend toward the blurring of entertainment/editorial and advertising content. Marketers are "embedding" products into the media so that consumers are not aware they are being advertised to. In radio, it is sometimes difficult to tell where DJ chit-chat ends and an advertisement begins.

Mobile Advertising

Mobile advertising is a form of advertising via mobile (wireless) phones or other mobile devices. It is a subset of mobile marketing.

It is estimated that mobile app install ads accounted for 30% of all mobile advertising revenue in 2014, and will top $4.6bn in 2015, and over $6.8bn by the end of 2019. Other ways mobile advertising can be purchased include working with a Mobile Demand Side Platform, in which ad impressions are bought in real-time on an Ad exchange.

Overview

Some see mobile advertising as closely related to online or internet advertising, though its reach is far greater — currently, most mobile advertising is targeted at mobile phones, that came at an estimated global total of $4.6bn as of 2009. Notably computers, including desktops and laptops, are currently estimated at 1.1 billion globally. Moreover, mobile advertising includes SMS and MMS advertising units in addition to the advertisement types served and processed via online channels.

It is probable that advertisers and media industry will increasingly take account of a bigger and fast-growing mobile market, though it remains at around 1% of global advertising spend. Mobile media is evolving rapidly and while mobile phones will continue to be the mainstay, it is not clear whether mobile phones based on cellular backhaul or smartphones based on WiFi hot spot or WiMAX hot zone will also strengthen. However, such is the emergence of this form of advertising, that there is now a dedicated global awards ceremony organised every year by Visiongain.

According to the research firm Berg Insight the global mobile advertising market that was estimated to €1 billion in 2008. Furthermore, Berg Insight forecasts the global mobile advertising market to grow at a compound annual growth rate of 43 percent to €8.7 billion in 2014.

Types of Mobile Ads

Mobile Rich Media

There are limitations to rich media on mobile because all of the coding must be done in HTML5, since the iOS does not support flash.

Handsets Display and Corresponding Ad Images

There are hundreds of handsets in the market and they differ by screen size and supported technologies (e.g. MMS, WAP 2.0). For color images, formats such as PNG, JPEG, GIF and BMP are typically supported, along with the monochrome WBMP format. The following gives an overview of various handset screen sizes and a recommended image size for each type.

Handset	Approx Handset Screen Size (px W x H)	Example Handsets	Ad Unit	Ad Size (pixels)
X-Large	320 x 320	Palm Treo 700P, Nokia E70	X-Large	300 x 50
Large	240 x 320	Samsung MM-A900, LG VX-8500 Chocolate, Sony Ericsson W910i	Large	216 x 36
Medium	176 x 208	Motorola RAZR, LG VX-8000, Motorola ROKR E1	Medium	168 x 28
Small	128 x 160	Motorola V195	Small	120 x 20

Source: Mobile Marketing Association

History

Martin Cooper invented a portable handset in 1973, when he was a project manager at Motorola. It was almost three decades after the idea of cellular communications was introduced by Bell Laboratories. Two decades later, cellular phones made a commercial debut in the mass market in the early 1990s. In the early days of cellular handsets, phone functionality was limited to dialing, and voice input/output.

When the second generation of mobile telecoms (so-called 2G) was introduced in Finland by Radiolinja (now Elisa) on the GSM standard (now the world's most common mobile technology with over 2 billion users) in 1991, the digital technology introduced data services. SMS text messaging was the first such service. The first person-to-person SMS text message was sent in Finland in December 1994. SMS (Short Message Service) gradually began to grow, becoming the largest data service by number of users in the world, currently with 74% of all mobile subscribers or 2.4 billion people active users of SMS in 2007.

One advantage of SMS is that while even in conference, users are able to send and receive brief messages unobtrusively, while enjoying privacy. Even in such environments as in a restaurant, café, bank, travel agency office, and so on, the users can enjoy some privacy by sending/receiving brief text messages in an unobtrusive way.

It would take six years from the launch of SMS until the first case of advertising would appear on this new data media channel, when a Finnish news provider offered free news headlines via SMS, sponsored by advertising. This led to rapid experimentation in mobile advertising and mobile marketing, and the world's first conference to discuss mobile advertising was held in London in 2000, sponsored by the Wireless Marketing Association (which later merged into the Mobile Marketing Association). The first books to discuss mobile advertising were Ahonen's M-Profits and Haig's Mobile Marketing in 2002. Several major mobile operators around the world launched their own mobile advertising arms, like Aircross in South Korea, owned by the parents of SK Telecoms the biggest mobile operator, or like D2 Communications in Japan, the joint venture of Japan's largest mobile operator NTT DoCoMo and Dentsu, Japan's largest ad agency.

Mobile as Media

This unobtrusive two-way communications caught the attention of media industry and advertisers as well as cellphone makers and telecom operators. Eventually, SMS became a new media - called the "seventh mass media channel" by several media and mobile experts - and even more, it is a two-way mobile media, as opposed to one-way immobile media like radios, newspapers and TV. Besides, the immediacy of responsiveness in this two-way media is a new territory found for media industry and advertisers, who are eager to measure up market response immediately. Additionally, the possibility of fast delivery of the messages and the ubiquity of the technology (it does not require any additional functionality from the mobile phone, all devices available today are capable of receiving SMS), make it ideal for time- and location-sensitive advertising, such as customer loyalty offers (ex. shopping centres, large brand stores), SMS promotions of events, etc. To leverage this strength of SMS advertising, timely and reliable delivery of messages is paramount, which is guaranteed by some SMS gateway providers.

Mobile media has begun to draw more significant attention from media giants and advertising industry since the mid-2000s, based on a view that mobile media was to change the way advertisements were made, and that mobile devices can form a new media sector. Despite this, revenues are still a small fraction of the advertising industry as a whole but are most certainly on the rise. Informa reported that mobile advertising in 2007 was worth $2.2 billion. This is less than 0.5% of the approximately $450 billion global advertising industry.

Types of mobile advertising are expected to change rapidly. In other words, mobile technology will come up with a strong push for identifying newer and unheard-of mobile multimedia, with the result that subsequent media migration will greatly stimulate a consumer behavioral shift and establish a paradigm shift in mobile advertising. A major media migration is on, as desktop Internet evolves into mobile Internet. One typical case in point is Nielsen's buyout of Telephia.

However it should be kept in mind that the rapid change in the technology used by mobile advertisers can also have adverse effect to the number of consumers being reached by the mobile advertisements, due to technical limitations of their mobile devices. Because of that, campaigns that aim to achieve wide response or are targeting lower income groups might be better off relying on older, more widespread mobile advertising technologies, such as SMS.

Viral Marketing

As mobile is an interactive mass media similar to the internet, advertisers are eager to utilize and make use of viral marketing methods, by which one recipient of an advertisement on mobile, will forward that to a friend. This allows users to become part of the advertising experience. At the bare minimum mobile ads with viral abilities can become powerful interactive campaigns. At the extreme, they can become engagement marketing experiences. A key element of mobile marketing campaigns is the most influential member of any target audience or community, which is called the alpha user

Future of Digital Marketing

Privacy Concern

Advocates have raised the issue of privacy. Targeted mobile marketing requires customization of ad content to reach interested and relevant customers. To customize such behavioral personal data, user profiling, data mining and other behavior watch tools are employed, and privacy advocates warn that this may cause privacy infringement.

Some mobile carriers offer freebie or cheaper rate plans in exchange for SMS or other mobile ads. However, mobile TV and mobile search may override this privacy concern, as soon as they are implemented on a full-blown basis. In a naive way to override privacy concern, however, User's prior consent needs to be obtained through membership to join or User account to set up. Both mobile TV and mobile search may supersede the way of getting Users' prior consent through membership or User account because users are free to choose mobile TV channels or mobile search services on a voluntary basis.

Interactivity

Mobile devices aim to outgrow the domain of voice-intensive cellphones and to enter a new world of multimedia mobile devices, like laptops, PDA phones and smartphones. Unlike the conventional one-way media like TV, radio and newspaper, web media has enabled two-way traffic, thereby introducing a new phase of interactive advertising, regardless of whether static or mobile. This user-centric approach was noted at the 96th annual conference of Association of National Advertisers in 2006, which described "a need to replace decades worth of top-down marketing tactics with bottom-up, grass-roots approaches". Many use 2d bar codes to make offline print material more interactive with their mobile device. This has been proven to be successful in Japan, UK, Philippines and has been catching on in Northern America.

Mobile Device Issues

Coincidentally, however, mobile devices are encountering technological bottlenecks in terms of battery life, formats, and safety issues.

In a broad sense, mobile devices are categorically broken down into portable and stationary equipment. Technically, mobile devices are categorized as below:

- Handheld [portable]
- Laptop, including ultraportable [portable]
- Dashtop, including GPS navigation, satellite radio, and WiMAX-enabled dashtop mobile payment platforms[fixed on dashboards]

The battery life and safety issues will perhaps combine to eventually push mobile equipment's inroads into vehicle dashtops. However, satellite-based GPS navigation and satellite radio may already hit a snag because of their part-time usage and technological hierarchy. Put differently, people want more functions than GPS navigation and satellite radios. The trend indicates an ongoing convergence into all-in-one dashtop mobile devices incorporating GPS navigators, satellite radios, MP3 players, mobile TV, mobile Internet, MVDER (vehicle black box), driving safety monitors, smartphones and even video games.

Online Advertising

Online advertising, also called online marketing or Internet advertising or web advertising, is a form of marketing and advertising which uses the Internet to deliver promotional marketing messages to consumers. It includes email marketing, search engine marketing (SEM), social media marketing, many types of display advertising (including web banner advertising), and mobile advertising. Like other advertising media, online advertising frequently involves both a publisher, who integrates advertisements into its online content, and an advertiser, who provides the advertisements to be displayed on the publisher's content. Other potential participants include advertising agencies who help generate and place the ad copy, an ad server which technologically delivers the ad and tracks statistics, and advertising affiliates who do independent promotional work for the advertiser.

In 2011, Internet advertising revenues in the United States surpassed those of cable television and nearly exceeded those of broadcast television. In 2013, Internet advertising revenues in the United States totaled $42.8 billion, a 17% increase over the $36.57 billion in revenues in 2012. U.S. internet ad revenue hit a historic high of $20.1 billion for the first half of 2013, up 18% over the same period in 2012. Online advertising is widely used across virtually all industry sectors.

Many common online advertising practices are controversial and increasingly subject to regulation. Online ad revenues may not adequately replace other publishers' revenue streams. Declining ad revenue has led some publishers to hide their content behind paywalls.

History

In early days of the Internet, online advertising was mostly prohibited. For example, two of the predecessor networks to the Internet, ARPANET and NSFNet, had "acceptable use policies" that banned network "use for commercial activities by for-profit institutions". The NSFNet began phasing out its commercial use ban in 1991.

Email

The first widely publicized example of online advertising was conducted via electronic mail. On 3 May 1978, a marketer from DEC (Digital Equipment Corporation), Gary Thuerk, sent an email to most of the ARPANET's American west coast users, advertising an open house for a new model of a DEC computer. Despite the prevailing acceptable use policies, electronic mail marketing rapidly expanded and eventually became known as "spam."

The first known large-scale non-commercial spam message was sent on 18 January 1994 by an Andrews University system administrator, by cross-posting a religious message to all USENET newsgroups. Four months later, Laurence Canter and Martha Siegel, partners in a law firm, broadly promoted their legal services in a USENET posting titled "Green Card Lottery – Final One?" Canter and Siegel's Green Card USENET spam raised the profile of online advertising, stimulating widespread interest in advertising via both Usenet and traditional email. More recently, spam has evolved into a more industrial operation, where spammers use armies of virus-infected computers (botnets) to send spam remotely.

Display Ads

Online banner advertising began in the early 1990s as page owners sought additional revenue streams to support their content. Commercial online service Prodigy displayed banners at the bottom of the screen to promote Sears products. The first clickable web ad was sold by Global Network Navigator in 1993 to a Silicon Valley law firm. In 1994, web banner advertising became mainstream when HotWired, the online component of Wired Magazine, sold banner ads to AT&T and other companies. The first AT&T ad on HotWired had a 44% click-through rate, and instead of directing clickers to AT&T's website, the ad linked to an online tour of seven of the world's most acclaimed art museums.

Search Ads

GoTo.com (renamed Overture in 2001, and acquired by Yahoo! in 2003) created the first search advertising keyword auction in 1998. Google launched its "AdWords" search advertising program in 2000 and introduced quality-based ranking allocation in 2002, which sorts search advertisements by a combination of bid price and searchers' likeliness to click on the ads.

Recent Trends

More recently, companies have sought to merge their advertising messages into editorial content or valuable services. Examples include Red Bull's Red Bull Media House streaming Felix Baumgartner's jump from space online, Coca-Cola's online magazines, and Nike's free applications for performance tracking. Advertisers are also embracing social media and mobile advertising; mobile ad spending has grown 90% each year from 2010 to 2013.

Delivery Methods

Display Advertising

Display advertising conveys its advertising message visually using text, logos, animations, videos, photographs, or other graphics. Display advertisers frequently target users with particular traits to increase the ads' effect. Online advertisers (typically through their ad servers) often use cookies, which are unique identifiers of specific computers, to decide which ads to serve to a particular consumer. Cookies can track whether a user left a page without buying anything, so the advertiser can later retarget the user with ads from the site the user visited.

As advertisers collect data across multiple external websites about a user's online activity, they can create a detailed picture of the user's interests to deliver even more targeted advertising. This aggregation of data is called behavioral targeting. Advertisers can also target their audience by using contextual to deliver display ads related to the content of the web page where the ads appear. Retargeting, behavioral targeting, and contextual advertising all are designed to increase an advertiser's return on investment, or ROI, over untargeted ads.

Advertisers may also deliver ads based on a user's suspected geography through geotargeting. A user's IP address communicates some geographic information (at minimum, the user's country or general region). The geographic information from an IP can be supplemented and refined with other proxies or information to narrow the range of possible locations. For example, with mobile devices, advertisers can sometimes use a phone's GPS receiver or the location of nearby mobile towers. Cookies and other persistent data on a user's machine may provide help narrowing a user's location further.

Web Banner Advertising

Web banners or banner ads typically are graphical ads displayed within a web page. Many banner ads are delivered by a central ad server.

Banner ads can use rich media to incorporate video, audio, animations, buttons, forms, or other interactive elements using Java applets, HTML5, Adobe Flash, and other programs.

Frame Ad (Traditional Banner)

Frame ads were the first form of web banners. The colloquial usage of "banner ads" often refers to traditional frame ads. Website publishers incorporate frame ads by setting aside a particular space on the web page. The Interactive Advertising Bureau's Ad Unit Guidelines proposes standardized pixel dimensions for ad units.

Pop-ups/Pop-unders

A pop-up ad is displayed in a new web browser window that opens above a website visitor's initial browser window. A pop-under ad opens a new browser window under a website visitor's initial browser window. Pop-under ads and similar technologies are now advised against by online au-thorities such as Google, who state that they "do not condone this practice".

Floating Ad

A floating ad, or overlay ad, is a type of rich media advertisement that appears superimposed over the requested website's content. Floating ads may disappear or become less obtrusive after a preset time period.

Expanding Ad

An expanding ad is a rich media frame ad that changes dimensions upon a predefined condition, such as a preset amount of time a visitor spends on a webpage, the user's click on the ad, or the user's mouse movement over the ad. Expanding ads allow advertisers to fit more information into a restricted ad space.

Trick Banners

A trick banner is a banner ad where the ad copy imitates some screen element users commonly encounter, such as an operating system message or popular application message, to induce ad clicks. Trick banners typically do not mention the advertiser in the initial ad, and thus they are a form of bait-and-switch. Trick banners commonly attract a higher-than-average click-through rate, but tricked users may resent the advertiser for deceiving them.

News Feed Ads

"News Feed Ads", also called "Sponsored Stories", "Boosted Posts", typically exist on Social Media Platforms that offer a steady stream of information updates ("news feed") in regulated formats (i.e. in similar sized small boxes with a uniform style). Those advertisements are intertwined with non-promoted news that the users are reading through. Those advertisements can be of any content, such as promoting a website, a fan page, an app, or a product.

Some examples are: Facebook's "Sponsored Stories", LinkedIn's "Sponsored Updates", and Twitter's "Promoted Tweets".

This display ads format falls into its own category because unlike banner ads which are quite distinguishable, News Feed Ads' format blends well into non-paid news updates. This format of online advertisement yields much higher click-through rates than traditional display ads

Display Advertising Process Overview

The process by which online advertising is displayed can involve many parties. In the simplest case, the web site publisher selects and serves the ads. Publishers which operate their own advertising departments may use this method.

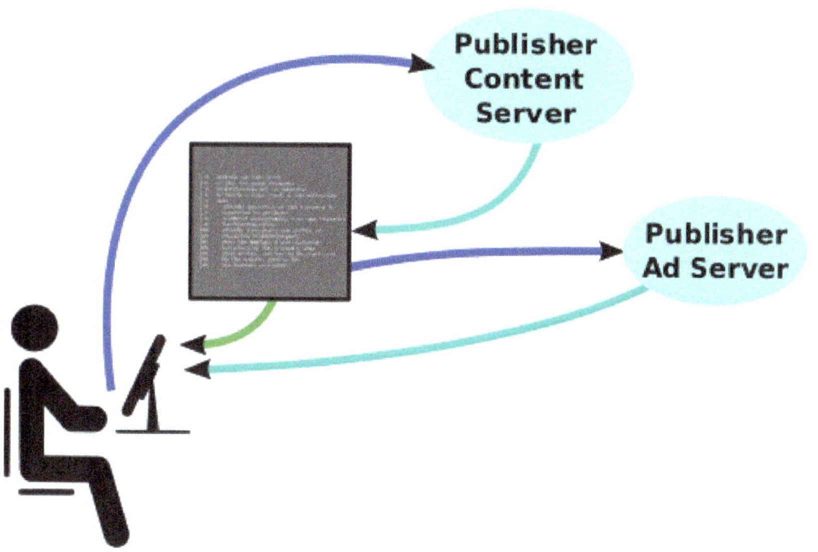

Online advertising serving process - simple publisher case

The ads may be outsourced to an advertising agency under contract with the publisher, and served from the advertising agency's servers.

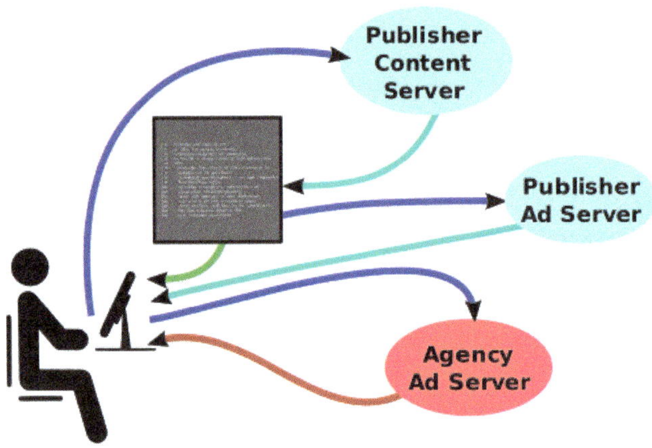

Online advertising serving process using an ad agency

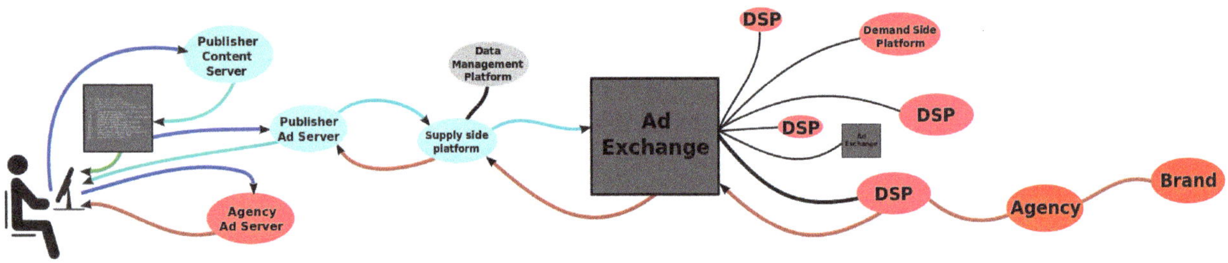

Online advertising serving process using online bidding

Alternatively, ad space may be offered for sale in a bidding market using an ad exchange and real-time bidding. This involves many parties interacting automatically in real time. In response to a request from the user's browser, the publisher content server sends the web page content to the user's browser over the Internet. The page does not yet contain ads, but contains links which cause the user's browser to connect to the publisher ad server to request that the spaces left for ads be filled in with ads. Information identifying the user, such as cookies and the page being viewed, is transmitted to the publisher ad server.

The publisher ad server then communicates with a supply-side platform server. The publisher is offering ad space for sale, so they are considered the supplier. The supply side platform also receives the user's identifying information, which it sends to a data management platform. At the data management platform, the user's identifying information is used to look up demographic information, previous purchases, and other information of interest to advertisers.

Broadly speaking, there are three types of data obtained through such a data management platform:

> *First party data* refers to the data retrieved from customer relationship management (CRM) platforms, in addition to website and paid media content or cross-platform data. This can include data from customer behaviors, actions or interests.
>
> *Second party data* refers to an amalgamation of statistics related to cookie pools on external publications and platforms. The data is provided directly from the source (adservers, hosted solutions for social or an analytics platform). It is also possible to negotiate a deal with a particular publisher to secure specific data points or audiences.

Third party data is sourced from external providers and often aggregated from numerous websites. Businesses sell third-party data and are able to share this via an array of distribution avenues.

This customer information is combined and returned to the supply side platform, which can now package up the offer of ad space along with information about the user who will view it. The supply side platform sends that offer to an ad exchange.

The ad exchange puts the offer out for bid to demand-side platforms. Demand side platforms act on behalf of ad agencies, who sell ads which advertise brands. Demand side platforms thus have ads ready to display, and are searching for users to view them. Bidders get the information about the user ready to view the ad, and decide, based on that information, how much to offer to buy the ad space. According to the Internet Advertising Bureau, a demand side platform has 10 milliseconds to respond to an offer. The ad exchange picks the winning bid and informs both parties.

The ad exchange then passes the link to the ad back through the supply side platform and the publisher's ad server to the user's browser, which then requests the ad content from the agency's ad server. The ad agency can thus confirm that the ad was delivered to the browser.

This is simplified, according to the IAB. Exchanges may try to unload unsold ("remnant") space at low prices through other exchanges. Some agencies maintain semi-permanent pre-cached bids with ad exchanges, and those may be examined before going out to additional demand side platforms for bids. The process for mobile advertising is different and may involve mobile carriers and handset software manufacturers.

Interstitial

An interstitial ad displays before a user can access requested content, sometimes while the user is waiting for the content to load. Interstitial ads are a form of interruption marketing.

Text Ads

A text ad displays text-based hyperlinks. Text-based ads may display separately from a web page's primary content, or they can be embedded by hyperlinking individual words or phrases to advertiser's websites. Text ads may also be delivered through email marketing or text message marketing. Text-based ads often render faster than graphical ads and can be harder for ad-blocking software to block.

Search Engine Marketing (SEM)

Search engine marketing, or SEM, is designed to increase a website's visibility in search engine results pages (SERPs). Search engines provide sponsored results and organic (non-sponsored) results based on a web searcher's query. Search engines often employ visual cues to differentiate sponsored results from organic results. Search engine marketing includes all of an advertiser's actions to make a website's listing more prominent for topical keywords.

Search Engine Optimization (SEO)

Search engine optimization, or SEO, attempts to improve a website's organic search rankings in

SERPs by increasing the website content's relevance to search terms. Search engines regularly update their algorithms to penalize poor quality sites that try to game their rankings, making optimization a moving target for advertisers. Many vendors offer SEO services.

Sponsored Search

Sponsored search (also called sponsored links, search ads, or paid search) allows advertisers to be included in the sponsored results of a search for selected keywords. Search ads are often sold via real-time auctions, where advertisers bid on keywords. In addition to setting a maximum price per keyword, bids may include time, language, geographical, and other constraints. Search engines originally sold listings in order of highest bids. Modern search engines rank sponsored listings based on a combination of bid price, expected click-through rate, keyword relevancy and site quality.

Social Media Marketing

Social media marketing is commercial promotion conducted through social media websites. Many companies promote their products by posting frequent updates and providing special offers through their social media profiles.

Mobile Advertising

Mobile advertising is ad copy delivered through wireless mobile devices such as smartphones, feature phones, or tablet computers. Mobile advertising may take the form of static or rich media display ads, SMS (Short Message Service) or MMS (Multimedia Messaging Service) ads, mobile search ads, advertising within mobile websites, or ads within mobile applications or games (such as interstitial ads, "advergaming," or application sponsorship). Industry groups such as the Mobile Marketing Association have attempted to standardize mobile ad unit specifications, similar to the IAB's efforts for general online advertising.

Mobile advertising is growing rapidly for several reasons. There are more mobile devices in the field, connectivity speeds have improved (which, among other things, allows for richer media ads to be served quickly), screen resolutions have advanced, mobile publishers are becoming more sophisticated about incorporating ads, and consumers are using mobile devices more extensively. The Interactive Advertising Bureau predicts continued growth in mobile advertising with the adoption of location-based targeting and other technological features not available or relevant on personal computers. In July 2014 Facebook reported advertising revenue for the June 2014 quarter of $2.68 billion, an increase of 67 percent over the second quarter of 2013. Of that, mobile advertising revenue accounted for around 62 percent, an increase of 41 percent on the previous year.

Email Advertising

Email advertising is ad copy comprising an entire email or a portion of an email message. Email marketing may be unsolicited, in which case the sender may give the recipient an option to opt out of future emails, or it may be sent with the recipient's prior consent (opt-in).

Chat Advertising

As opposed to static messaging, chat advertising refers to real time messages dropped to users on certain sites. This is done by the usage of live chat software or tracking applications installed within certain websites with the operating personnel behind the site often dropping adverts on the traffic surfing around the sites. In reality this is a subset of the email advertising but different because of its time window.

Online Classified Advertising

Online classified advertising is advertising posted online in a categorical listing of specific products or services. Examples include online job boards, online real estate listings, automotive listings, online yellow pages, and online auction-based listings. Craigslist and eBay are two prominent providers of online classified listings.

Adware

Adware is software that, once installed, automatically displays advertisements on a user's computer. The ads may appear in the software itself, integrated into web pages visited by the user, or in pop-ups/pop-unders. Adware installed without the user's permission is a type of malware.

Affiliate Marketing

Affiliate marketing (sometimes called lead generation) occurs when advertisers organize third parties to generate potential customers for them. Third-party affiliates receive payment based on sales generated through their promotion. Affiliate marketers generate traffic to offers from affiliate networks, and when the desired action is taken by the visitor, the affiliate earns a commission. These desired actions can be an email submission, a phone call, filling out an online form, or an online order being completed.

Content Marketing

Content marketing is any marketing that involves the creation and sharing of media and publishing content in order to acquire and retain customers. This information can be presented in a variety of formats, including blogs, news, video, white papers, e-books, infographics, case studies, how-to guides and more.

Considering that most marketing involves some form of published media, it is almost (though not entirely) redundant to call 'content marketing' anything other than simply 'marketing'. There are, of course, other forms of marketing (in-person marketing, telephone-based marketing, word of mouth marketing, etc.) where the label is more useful for identifying the type of marketing. However, even these are usually merely presenting content that they are marketing as information in a way that is different from traditional print, radio, TV, film, email, or web media.

Online Marketing Platform

Online marketing platform (OMP) is an integrated web-based platform that combines the benefits of a business directory, local search engine, search engine optimisation (SEO) tool, customer relationship management (CRM) package and content management system (CMS). Ebay and Amazon are used as online marketing and logistics management platforms. On Facebook, Twitter, YouTube, Pinterest, LinkedIn, and other Social Media, retail online marketing is also used. Online business marketing platforms such as Marketo, Aprimo, MarketBright and Pardot have been bought by major IT companies (Eloqua-Oracle, Neolane-Adobe and Unica-IBM).

Compensation Methods

Advertisers and publishers use a wide range of payment calculation methods. In 2012, advertisers calculated 32% of online advertising transactions on a cost-per-impression basis, 66% on customer performance (e.g. cost per click or cost per acquisition), and 2% on hybrids of impression and performance methods.

CPM (Cost Per Mille)

Cost per mille, often abbreviated to CPM, means that advertisers pay for every thousand displays of their message to potential customers (mille is the Latin word for thousand). In the online context, ad displays are usually called "impressions." Definitions of an "impression" vary among publishers, and some impressions may not be charged because they don't represent a new exposure to an actual customer. Advertisers can use technologies such as web bugs to verify if an impression is actually delivered.

Publishers use a variety of techniques to increase page views, such as dividing content across multiple pages, repurposing someone else's content, using sensational titles, or publishing tabloid or sexual content.

CPM advertising is susceptible to "impression fraud," and advertisers who want visitors to their sites may not find per-impression payments a good proxy for the results they desire.

CPC (Cost Per Click)

CPC (Cost Per Click) or PPC (Pay per click) means advertisers pay each time a user clicks on the ad. CPC advertising works well when advertisers want visitors to their sites, but it's a less accurate measurement for advertisers looking to build brand awareness. CPC's market share has grown each year since its introduction, eclipsing CPM to dominate two-thirds of all online advertising compensation methods.

Like impressions, not all recorded clicks are valuable to advertisers. GoldSpot Media reported that up to 50% of clicks on static mobile banner ads are accidental and resulted in redirected visitors leaving the new site immediately.

CPE (Cost Per Engagement)

Cost per engagement aims to track not just that an ad unit loaded on the page (i.e., an impression was served), but also that the viewer actually saw and/or interacted with the ad.

CPV (Cost Per View)

Cost per view video advertising. Both Google and TubeMogul endorsed this standardized CPV metric to the IAB's (Interactive Advertising Bureau) Digital Video Committee, and it's garnering a notable amount of industry support. CPV is the primary benchmark used in YouTube Advertising Campaigns, as part of Google's AdWords platform.

CPI (Cost Per Install)

The CPI compensation method is specific to mobile applications and mobile advertising. In CPI ad campaigns brands are charged a fixed of bid rate only when the application was installed.

Attribution of Ad Value

In marketing, "attribution" is the measurement of effectiveness of particular ads in a consumer's ultimate decision to purchase. Multiple ad impressions may lead to a consumer "click" or other action. A single action may lead to revenue being paid to multiple ad space sellers.

Other Performance-based Compensation

CPA (Cost Per Action or Cost Per Acquisition) or PPP (Pay Per Performance) advertising means the advertiser pays for the number of users who perform a desired activity, such as completing a purchase or filling out a registration form. Performance-based compensation can also incorporate revenue sharing, where publishers earn a percentage of the advertiser's profits made as a result of the ad. Performance-based compensation shifts the risk of failed advertising onto publishers.

Fixed Cost

Fixed cost compensation means advertisers pay a fixed cost for delivery of ads online, usually over a specified time period, irrespective of the ad's visibility or users' response to it. One examples is CPD (cost per day) where advertisers pay a fixed cost for publishing an ad for a day irrespective of impressions served or clicks.

Benefits of Online Advertising

Cost

The low costs of electronic communication reduce the cost of displaying online advertisements compared to offline ads. Online advertising, and in particular social media, provides a low-cost means for advertisers to engage with large established communities. Advertising online offers better returns than in other media.

Measurability

Online advertisers can collect data on their ads' effectiveness, such as the size of the potential audience or actual audience response, how a visitor reached their advertisement, whether the advertisement resulted in a sale, and whether an ad actually loaded within a visitor's view. This helps online advertisers improve their ad campaigns over time.

Formatting

Advertisers have a wide variety of ways of presenting their promotional messages, including the ability to convey images, video, audio, and links. Unlike many offline ads, online ads also can be interactive. For example, some ads let users input queries or let users follow the advertiser on social media. Online ads can even incorporate games.

Targeting

Publishers can offer advertisers the ability to reach customizable and narrow market segments for targeted advertising. Online advertising may use geo-targeting to display relevant advertisements to the user's geography. Advertisers can customize each individual ad to a particular user based on the user's previous preferences. Advertisers can also track whether a visitor has already seen a particular ad in order to reduce unwanted repetitious exposures and provide adequate time gaps between exposures.

Coverage

Online advertising can reach nearly every global market, and online advertising influences offline sales.

Speed

Once ad design is complete, online ads can be deployed immediately. The delivery of online ads does not need to be linked to the publisher's publication schedule. Furthermore, online advertisers can modify or replace ad copy more rapidly than their offline counterparts.

Concerns

Security Concerns

According to a US Senate investigation, the current state of online advertising endangers the security and privacy of users.

Banner Blindness

Eye-tracking studies have shown that Internet users often ignore web page zones likely to contain display ads (sometimes called "banner blindness"), and this problem is worse online than in offline media. On the other hand, studies suggest that even those ads "ignored" by the users may influence the user subconsciously.

Fraud on the Advertiser

There are numerous ways that advertisers can be overcharged for their advertising. For example, click fraud occurs when a publisher or third parties click (manually or through automated means) on a CPC ad with no legitimate buying intent. For example, click fraud can occur when a competitor clicks on ads to deplete its rival's advertising budget, or when publishers attempt to manufacture revenue.

Click fraud is especially associated with pornography sites. In 2011, certain scamming porn websites launched dozens of hidden pages on each visitor's computer, forcing the visitor's computer to click on hundreds of paid links without the visitor's knowledge.

As with offline publications, online impression fraud can occur when publishers overstate the number of ad impressions they have delivered to their advertisers. To combat impression fraud, several publishing and advertising industry associations are developing ways to count online impressions credibly.

Technological Variations

Heterogeneous Clients

Because users have different operating systems, web browsers and computer hardware (including mobile devices and different screen sizes), online ads may appear to users differently from how the advertiser intended, or the ads may not display properly at all. A 2012 comScore study revealed that, on average, 31% of ads were not "in-view" when rendered, meaning they never had an opportunity to be seen. Rich media ads create even greater compatibility problems, as some developers may use competing (and exclusive) software to render the ads.

Furthermore, advertisers may encounter legal problems if legally required information doesn't actually display to users, even if that failure is due to technological heterogeneity. In the United States, the FTC has released a set of guidelines indicating that it's the advertisers' responsibility to ensure the ads display any required disclosures or disclaimers, irrespective of the users' technology.

Ad-blocking

Ad-blocking, or ad filtering, means the ads do not appear to the user because the user uses technology to screen out ads. Many browsers block unsolicited pop-up ads by default. Other software programs or browser add-ons may also block the loading of ads, or block elements on a page with behaviors characteristic of ads (e.g. HTML autoplay of both audio and video). Approximately 9% of all online page views come from browsers with ad-blocking software installed, and some publishers have 40%+ of their visitors using ad-blockers.

Anti-targeting Technologies

Some web browsers offer privacy modes where users can hide information about themselves from publishers and advertisers. Among other consequences, advertisers can't use cookies to serve targeted ads to private browsers. Most major browsers have incorporated Do Not Track options into their browser headers, but the regulations currently are only enforced by the honor system.

Privacy Concerns

The collection of user information by publishers and advertisers has raised consumer concerns about their privacy. Sixty percent of Internet users would use Do Not Track technology to block all collection of information if given an opportunity. Over half of all Google and Facebook users are concerned about their privacy when using Google and Facebook, according to *Gallup*.

Many consumers have reservations about online behavioral targeting. By tracking users' online activities, advertisers are able to understand consumers quite well. Advertisers often use technology, such as web bugs and respawning cookies, to maximizing their abilities to track consumers. According to a 2011 survey conducted by Harris Interactive, over half of Internet users had a negative impression of online behavioral advertising, and forty percent feared that their personally-identifiable information had been shared with advertisers without their consent. Consumers can be especially troubled by advertisers targeting them based on sensitive information, such as financial or health status.

Trustworthiness of Advertisers

Scammers can take advantage of consumers' difficulties verifying an online persona's identity, leading to artifices like phishing (where scam emails look identical to those from a well-known brand owner) and confidence schemes like the Nigerian "419" scam. The Internet Crime Complaint Center received 289,874 complaints in 2012, totaling over half a billion dollars in losses, most of which originated with scam ads.

Consumers also face malware risks, i.e. malvertising, when interacting with online advertising. Cisco's 2013 Annual Security Report revealed that clicking on ads was 182 times more likely to install a virus on a user's computer than surfing the Internet for porn. For example, in August 2014 Yahoo's advertising network reportedly saw cases of infection of a variant of Cryptolocker ransomware.

Spam

The Internet's low cost of disseminating advertising contributes to spam, especially by large-scale spammers. Numerous efforts have been undertaken to combat spam, ranging from blacklists to regulatorily-required labeling to content filters, but most of those efforts have adverse collateral effects, such as mistaken filtering.

Regulation

In general, consumer protection laws apply equally to online and offline activities.[i] However, there are questions over which jurisdiction's laws apply and which regulatory agencies have enforcement authority over transborder activity.

As with offline advertising, industry participants have undertaken numerous efforts to self-regulate and develop industry standards or codes of conduct. Several United States advertising industry organizations jointly published *Self-Regulatory Principles for Online Behavioral Advertising* based on standards proposed by the FTC in 2009. European ad associations published a similar document in 2011. Primary tenets of both documents include consumer control of data transfer to third parties, data security, and consent for collection of certain health and financial data. Neither framework, however, penalizes violators of the codes of conduct.

Privacy and Data Collection

Privacy regulation can require users' consent before an advertiser can track the user or communicate with the user. However, affirmative consent ("opt in") can be difficult and expensive to obtain. Industry participants often prefer other regulatory schemes.

Different jurisdictions have taken different approaches to privacy issues with advertising. The United States has specific restrictions on online tracking of children in the Children's Online Privacy Protection Act (COPPA), and the FTC has recently expanded its interpretation of COPPA to include requiring ad networks to obtain parental consent before knowingly tracking kids. Otherwise, the U.S. Federal Trade Commission frequently supports industry self-regulation, although increasingly it has been undertaking enforcement actions related to online privacy and security. The FTC has also been pushing for industry consensus about possible Do Not Track legislation.

In contrast, the European Union's "Privacy and Electronic Communications Directive" restricts websites' ability to use consumer data much more comprehensively. The EU limitations restrict targeting by online advertisers; researchers have estimated online advertising effectiveness decreases on average by around 65% in Europe relative to the rest of the world.

Delivery Methods

Many laws specifically regulate the ways online ads are delivered. For example, online advertising delivered via email is more regulated than the same ad content delivered via banner ads. Among other restrictions, the U.S. CAN-SPAM Act of 2003 requires that any commercial email provide an opt-out mechanism. Similarly, mobile advertising is governed by the Telephone Consumer Protection Act of 1991 (TCPA), which (among other restrictions) requires user opt-in before sending advertising via text messaging.

References

- Moe, Wendy W. (2013). "Chapter 9: Targeting Display Advertising". Advanced Database Marketing: Innovative Methodologies & Applications for Managing Customer Relationships (PDF). Gower Publishing, London. ISBN 978-1409444619. Retrieved 20 June 2013.
- Tulloch, Mitch (2003). Koch, Jeff; Haynes, Sandra, eds. Microsoft Encyclopedia of Security. Redmond, Washington: Microsoft Press. p. 16. ISBN 0-7356-1877-1.
- Pieters, Rik (2008). "A Review of Eye-Tracking Research in Marketing". Review of Marketing Research. 4: 123–147. doi:10.1108/s1548-6435(2008)0000004009. ISBN 978-0-7656-2092-7.
- Kim Ji Young (11 October 2011). "Hallyu Stars Earning Big Money Through Overseas CFs". enewsWorld. CJ E&M. Retrieved 2016-03-25.
- De Atley, Richard (1985-09-07). "VCRs put entertainment industry into fast-forward frenzy". The Free Lance-Star. Associated Press. pp. 12–TV. Retrieved 25 January 2015.
- Hoelzel, Mark. "The mobile app-install ad is driving a boom in mobile ad spend — and not just among game makers". Business Insider. Business Insider. Retrieved 23 May 2015.
- Limon, Kathleen. "THE DIFFERENCE BETWEEN FIRST, SECOND, AND THIRD PARTY DATA AND HOW TO USE THEM". retargeter.com. Retargeter. Retrieved 3 April 2015.
- "Marketers Focus on Making Attribution Data Actionable: Industry experts discuss real-time optimisation of cross-platform attribution findings". emarketer.com. Retrieved 7 September 2015.
- Lucian Constantin (2014-05-15). "Online advertising poses significant security, privacy risks to users, US Senate report says". Network World. Retrieved 2015-12-24.
- "IAB internet advertising revenue report 2013 first six months' results" (PDF). Internet Advertising Bureau. October 2013. Retrieved 4 March 2014.
- "IAB internet advertising revenue report: 2012 full year results" (PDF). PricewaterhouseCoopers, Internet Ad-

vertising Bureau. April 2013. Retrieved 12 June 2013.

- "NSFNet Acceptable Use Policy". Information Policies: A Compilation of Position Statements, Principles, Statutes, and Other Pertinent Statements. Coalition for Networked Information. Retrieved 24 June 2013.
- "Junk Mail". Electronic Billboards on the Digital Superhighway: A Report of the Working Group on Internet Advertising. The Coalition for Networked Information. 28 September 1994. Retrieved 24 June 2013.
- "Google Introduces New Pricing for Popular Self-Service Online Advertising Program". Google. February 2002. Retrieved June 2013.
- Steel, Emily; Angwin, Julia (3 August 2010). "On the Web's Cutting Edge, Anonymity in Name Only". The Wall Street Journal. Retrieved 10 June 2013.
- Hanley, M.; Becker, M. (2007). "Cell Phone Usage and Advertising Acceptance Among College Students: A Four-Year Analysis". 2008 AEJMC Conference: Advertising Division–Research. Retrieved 13 June 2013.
- Kuckera, Ben (17 April 2013). "Ad-blockers, the games press, and why sexy cosplay galleries lead to better reporting". Penny Arcade Report. Archived from the original on 8 June 2013. Retrieved 14 June 2013.

Production and Broadcasting

7

Production is an integral part of broadcasting and this chapter discusses production terminology like broadcast network, broadcast automation, outside broadcasting, electronic field production, electronic news-gathering, broadcast-safe, broadcast delay and bleep censor. The reader is informed about the production side of broadcasting and its importance.

Broadcast Network

A broadcast network is an organization, such as a corporation or voluntary association, that provides live or recorded content (such as movies, newscasts, sports, public affairs programming, and other programs) for broadcast over a group of radio or television stations. Most networks are primarily either a television network or a radio network, although some organizations run both types of networks.

Logos of the five major U.S. broadcast networks (clockwise from top left: NBC, CBS, ABC, The CW, and FOX.)

Streaming media, Internet radio, and webcasting are sometimes considered forms of broadcasting despite the lack of broadcast stations; in which case its practitioners may also be called "broadcasters" or even "broadcast networks".

History

American Networks

Reginald Fessenden, a former engineer and communications researcher for the U.S. Weather Bureau was the first individual to transmit a regular radio broadcast. His broadcasts were to ships at sea, for which he used radio telegraphy equipment. His programs consisted of a recorded Handel piece, a violin performance, and a reading from the Bible. He claimed to be the first to transmit the human voice. General Electric was encouraged years later to create the Radio Corporation of America (RCA). Around this time, AT&T also became involved in radio.

AT&T

With all of these stations going on the air, AT&T devised the concept of sharing programming to save on effort required to create programming. In addition, programs sometimes appealed to the audiences of more than one station. When two or three stations link together from a telephone line it is called chain broadcasting. Since the phone lines were owned by AT&T, they were the first corporation to begin the sharing of two or three stations via telephone lines.

In 1924, the *Eveready Hour* was broadcast over 12 stations, many of them owned by AT&T. *Eveready Hour* was the first commercially sponsored variety show in the history of broadcasting. National companies were able to reach large portions of the nation with their brand names and slogans in a resourceful manner. By 1925 AT&T had linked together 26 stations in a network. They also ran an owned & operated (O&O) station, WEAFNinew York.

RCA

Radio Corporation of America (RCA) followed suit, using AT&T's network model. However, conflict resulted as RCA needed to lease phone lines from AT&T. The Federal Trade Commission (FTC) took notice of this and charged AT&T with United States antitrust law violations. In 1926 AT&T sold off their broadcasting interests to RCA. RCA agreed to lease network connections from AT&T and for several decades AT&T made a profitable business in radio and television networking.

NBC

In 1926, RCA created the National Broadcasting Company (NBC) under David Sarnoff. When NBC first launched, it comprised two groups of stations that worked together, with different programs and sponsors to support. The two networks were called NBC Red and NBC Blue. NBC Red covered stronger stations and NBC Blue covered weaker stations.

In 1941, the Federal Communications Commission (FCC) issued the Report on Chain Broadcasting, an investigation into the monopolistic practices of the radio networks. The FCC cited NBC Red and NBC Blue as a primary concern, which the commission thought was anti-competitive. Since the FCC does not have the power to regulate the networks, they decided aim the regulations to the stations. In 1943, the Supreme Court of the United States took away the FCC's power to enforce chain broadcasting regulations. NBC Blue was sold to Edward Noble who later named it the American Broadcasting Company (ABC). NBC also had a chain of shortwave stations, called NBC White Network, in the 1930s.

CBS

During 1926 and 1927 home radios were increasing dramatically in popularity. The third radio network that came about was the Columbia Broadcasting System (CBS). William S. Paley bought Columbia. CBS struggled at first because both NBC networks had a head start. Paley helped out CBS to get them where NBC was. Under Paley's leadership, CBS focused on entertainment programming, news, and news affiliation. Regional networks on CBS also existed in various parts of the country. CBS later hired Edward R. Murrow who is credited with boosting ratings dramatically. Murrow and CBS covered the war in Europe while Adolf Hitler was in power. NBC and ABC withdrew from the war for safety reasons. As a result of taking the risk, CBS's ratings skyrocketed.

ABC

NBC Blue was sold to Edward Noble, and became the American Broadcasting Company (ABC). ABC was born due to the controversy that the FCC had with NBC Blue and NBC Red. By the mid-1940s the broadcasting arena was a big Three television networks battle. Noble's network almost went bankrupt and in 1951 Leonard Goldenson and United Paramount Theaters bought ABC for $25 million. In 1964, ABC won the ratings race in the fifty largest U.S markets. In the 1970–71 season, ABC ranked #1 in the Nielsen ratings with a medical drama called Marcus Welby, M.D., the first ABC television show to top the list.

Canadian National Networks

CBC

Canadian Broadcasting Corporation

CTV

CTV Television Network

Global

Global Television Network

Broadcast Automation

Broadcast automation incorporates the use of broadcast programming technology to automate broadcasting operations. Used either at a broadcast network, radio station or a television station, it can run a facility in the absence of a human operator. They can also run in a "live assist" mode when there are on-air personnel present at the master control, television studio or control room.

The radio transmitter end of the airchain is handled by a separate automatic transmission system (ATS).

History

Originally, in the USA, many (if not most) broadcast licensing authorities required a licensed board operator to run every station at all times, meaning that every DJ had to pass an exam to obtain a license to be on-air, if their duties also required them to ensure proper operation of the transmitter. This was often the case on overnight and weekend shifts when there was no broadcast engineer present, and all of the time for small stations with only a contract engineer on call.

In the U.S., it was also necessary to have an operator on duty at all times in case the Emergency Broadcast System (EBS) was used, as this had to be triggered manually. While there has not been a requirement to relay any other warnings, any mandatory messages from the U.S. president would have had to first be authenticated with a code word sealed in a pink envelope sent annually to stations by the Federal Communications Commission (FCC).

Gradually, the quality and reliability of electronic equipment improved, regulations were relaxed, and no operator had to be present (or even available) while a station was operating. In the U.S., this came about when the EAS replaced the EBS, starting the movement toward automation to assist, and sometimes take the place of, the live disc jockeys (DJs) and radio personalities.

Early Analog Systems

Harris automation system used at the former WWJQ (now WPNW) in 1993.

Early automation systems were electromechanical systems which used relays. Later systems were "computerized" only to the point of maintaining a schedule, and were limited to radio rather than TV. Music would be stored on reel-to-reel audio tape. Subaudible tones on the tape marked the end of each song. The computer would simply rotate among the tape players until the computer's internal clock matched that of a scheduled event. When a scheduled event would be encountered, the computer would finish the currently-playing song and then execute the scheduled block of events. These events were usually advertisements, but could also include the station's top-of-hour station identification, news, or a bumper promoting the station or its other shows. At the end of the block, the rotation among tapes resumed.

Advertisements, jingles, and the top-of-hour station identification required by law were often on "carts". Short for cartridges, these were endless tapes similar to 8-track tapes, and looked nearly identical as well. A primary difference between carts and 8 track deals with the pinch roller and capstan. The roller was self-contained in an 8-track; carts had a slot for a pinch roller on a spindle which was activated by solenoid upon pressing the start button on the cart machine. This allowed for nearly instantaneous playback start without artifacts. Mechanical carousels would rotate the carts in and out of multiple tape players as dictated by the computer. Time announcements were provided by a pair of dedicated cart players, with the even minutes stored on one and the odd minutes on the other. This meant an announcement would always be ready to play, even if the minute was changing when the announcement was triggered. The system did require attention throughout the day to change reels as they ran out and reload carts. It became obsolete when a method was developed to automatically rewind and re-cue the reel tapes when they ran out, extending 'walk-away' time indefinitely.

Radio station WIRX may have been one of the world's first completely automated radio stations, built and designed by Brian Jeffrey Brown in 1963 when Brown was only 10 years old. The station broadcast in a classical format, called "More Good Music (MGM)" and featured five-minute bottom-of-the-hour news feeds from the Mutual Broadcasting System. The heart of the automation was an 8 x 24 telephone stepping relay which controlled two reel-to-reel tape decks, one twelve inch Ampex machine providing the main program audio and a second RCA seven inch machine providing "fill" music. The tapes played by these machines were originally produced in the MWF's Madison, Wisconsin production facility by WSJM Chief Engineer Richard E. McLemore (and later in-house at WSJM) with sub-audible tones used to signal the end of a song. The stepping relay was programmed by slide switches in the front of the two relay racks which housed the equipment. The news feeds were triggered by a microswitch which was attached to a Western Union clock and tripped by the minute hand of the clock. and then reset the stepping relay. Originally, 30-minute station identification was accomplished by a simulcast switch in the control booth for sister station WSJM (AM), whereupon the disc jockey in the booth would announce "This is WSJM-AM and... (then pressing the momentary contact button) ...WSJM-FM, St. Joseph, Michigan." This only lasted about six months, however, and a standard tape cartridge player was wired in to announce the station identification and triggered by the Western Union clock.

Solidyne GMS200 tape recorder with computer self-adjustment. Argentina 1980-1990

A different technology appeared in 1980 with the analog recorders made by Solidyne, which used a computer-controlled tape positioning system. Four GMS 204 units were controlled from a 6809 microprocessor, with the program stored in a solid-state plug-in memory module. This system has a limited programming time of about eight hours.

Satellite programming often used audible dual-tone multi-frequency (DTMF) signals to trigger events at affiliate stations. This allowed the automatic local insertion of ads and station IDs. Because there are 12 (or 16) tone pairs, and typically four tones were sent in rapid succession (less than one second), more events could be triggered than by sub-audible tones (usually 25 Hz and 35 Hz).

Modern Digital Systems

Modern systems run on hard disk, where all of the music, jingles, advertisements, voice tracks, and other announcements are stored. These audio files may be either compressed or uncompressed, or often with only minimal compression as a compromise between file size and quality. For radio software, these disks are usually in computers, sometimes running their own custom operating systems, but more often running as an application on a stable OS like GNU/Linux, Windows NT or others.

Scheduling was an important advance of these systems, allowing for exact timing. Some systems use GPS satellite receivers to obtain exact atomic time, for perfect synchronization with satellite-delivered programming. Reasonably-accurate timekeeping can also be obtained with the use of Internet protocols (IP) like Network Time Protocol (NTP).

Automation systems are also more interactive than ever before with digital audio workstation (DAW) with console automation and can even record from a telephone hybrid to play back an edited conversation with a telephone caller. This is part of a system's live-assist mode.

The use of automation software and voice tracks to replace live DJs is a current trend in radio broadcasting, done by many Internet radio and adult hits stations. Stations can even be voice-tracked from another city far away, now often delivering sound files over the Internet. In the U.S., this is a common practice under controversy for making radio more generic and artificial. Having local content is also touted as a way for traditional stations to compete with satellite radio, where there may be no radio personality on the air at all.

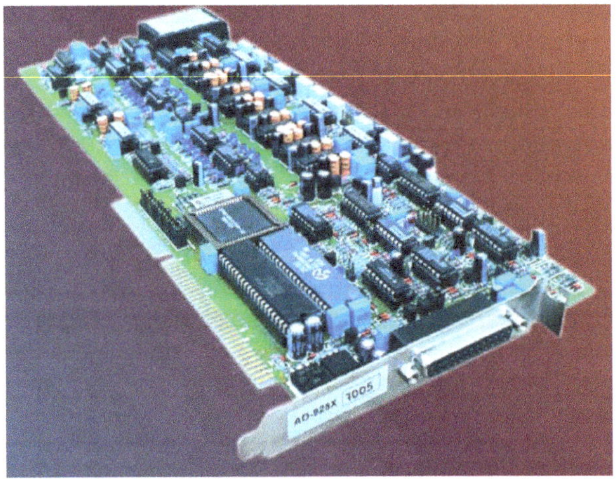

Solidyne 922: The first bit compression card for PC, 1990

A commercially available, for-sale product named Audicom was introduced by Oscar Bonello in 1989. It is based on psychoacoustic lossy compression, the same principle being used in most modern lossy audio encoders such as MP3 and Advanced Audio Coding (AAC), and it allowed both broadcast automation and recording to hard drives.

Television

In television, playout automation is also becoming more practical as the storage space of hard drives increases. Television shows and television commercials, as well as digital on-screen graphics (DOG or BUG), can all be stored on video servers remotely controlled by computers utilizing the 9-Pin Protocol and the Video Disk Control Protocol (VDCP). These systems can be very extensive, tied-in with parts that allow the "ingest" (as it is called in the industry) of video from satellite networks and electronic news gathering (ENG) operations and management of the video library, including archival of footage for later use. In ATSC, Programming Metadata Communication Protocol (PMCP) is then used to pass information about the video through the airchain to Program and System Information Protocol (PSIP), which transmits the current electronic program guide (EPG) information over digital television to the viewer.

Outside Broadcasting

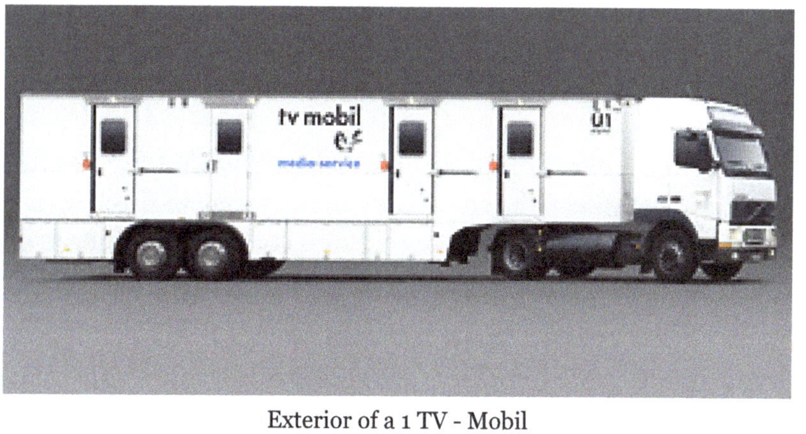

Exterior of a 1 TV - Mobil

Outside broadcasting (OB) is the electronic field production (EFP) of television or radio programmes (typically to cover television news and sports television events) from a mobile remote broadcast television studio. Professional video camera and microphone signals come into the production truck for processing, recording and possibly transmission. The mobile production control room (PCR) is known as a "production truck", "scanner" (a BBC term), "mobile unit", "remote truck", "live truck", "OB van" or "live eye". In the United States an "OB van" is smaller in size than a production truck and generally requires two or three people in the field to manage.

Modern Applications

In the past many outside broadcasting applications have relied on using satellite uplinks in order to broadcast live audio and video back to the studio. While this has its advantages such as the ability to set up virtually anywhere covered by the respective geostationary satellite, satellite uplinking is very expensive and the round trip latency is in the range of 240 to 280 milliseconds. Modern applications such as hardware and software IP codecs have allowed the use of public 3G/4G networks to broadcast video and audio. The latency of 3G is around 100–500 ms, while 4G is less than 100 ms.

Interior

A typical OB van is usually divided into five parts.

- Parts of the television crew are located in the first and largest part, the video production area. The television director, technical director, assistant director, character generator (CG) operator and television producers usually sit in front of a wall of video monitors. The technical director sits in front of the video switcher. The video monitors show all the video feeds from various sources, including computer graphics, professional video cameras, video tape recorder (VTR), video servers and slow-motion replay machines. The wall of monitors also contains a preview monitor showing what could be the next source on air (does not have to be depending on how the video switcher is set up) and a program monitor that shows the feed currently going to air or being recorded. The keyed *dirty feed* (with digital on-screen graphic) is what is actually transmitted back to the central studio that is controlling the outside broadcast. A clean feed (without the graphics) could be sent to other trucks for use in their production. The video switcher is usually operated by one person called the technical director (TD) and is responsible for switching the video sources to air as directed. Behind the directors there is usually a desk with monitors for the editors to operate. It is essential that the directors and editors are in communication with each other during events, so that replays and slow-motion shots can be selected and aired. The "production room" in most sporting events has a graphics operator and sometimes a font coordinator who are in charge of the graphics, statistics, and the showing of the names of commentators or the players to be shown on air. Most sports also have a "box operator" (or "bug box operator") who controls the graphic seen either on the bottom or top of the screen that shows the score as seen at home. These operators can also show on-air stats, control the clock, and many times are also in charge of showing sponsors during play.

- The second part of a van is where the audio engineer has an audio mixer (being fed with all the various audio feeds: reporters, commentary, on-field microphones, etc.). The audio engineer can control which channels are added to the output and follows instructions from the director. They relay the information from producers and directors to their A2's who typically set up the audio cables and equipment throughout the arenas and the booth where the commentators sit. The audio engineer normally also has a dirty feed monitor to help with the synchronization of sound and video.

- The third part of the truck is the VTR area. The tape area has a collection of machines including video servers and may also house additional power supplies or computer equipment. The "tape room" has EVS (or LSM) operators who have one or more cameras that go into their machines and can be played back for replays when an exciting or important play occurs during the game. EVS operators also play replay rollouts that lead into commercial breaks or show the highlights of the event at the end of play. These operators also can play back in slow motion or pause to show a key part of the action.

- The fourth part is the video control area where the professional video cameras are controlled using camera control units (CCU) by one or two operators, to make sure that the iris is at the correct exposure and that all the cameras look the same. These operators can shade, balance, and focus the cameras from this position inside the truck. This area is controlled by an operator called a V1 and depending on the size of the show and/or the broadcast company may have a V2.

- The fifth part is transmission, where the signal is monitored by and engineered for quality control purposes and is transmitted or sent to other trucks. The transmission is monitored by the truck engineers to ensure the people at home have a good picture and a high quality signal output.

Electronic Field Production

Director of Photography with Sony FS7 camera and Convergent Design Odyssey 7Q+ recorder monitor, a video production rig used for ENG and EFP shoots

Electronic field production (EFP) is a television industry term referring to a video production which takes place in the field, outside of a formal television studio, in a practical location or special venue. In the field-defining text, Television Production Handbook, now in its 12th edition, author Herbert Zettl defines EFP as using "both ENG and studio techniques. From ENG it borrows its mobility and flexiblity; from the studio it borrows its production care and quality control. EFP takes place on location (which may include shooting in someone's living room) and has to adapt to the location conditions... Good lighting and audio are always difficult to achieve in EFP, regardless of whether you are outdoors or indoors. Compared to ENG, in which you simply respond to a situation, EFP needs careful planning."

Multi-camera exterior EFP shoot at a BMX Rally, California, 2004

Typical applications of electronic field production include awards shows, concerts, major interviews for newsmagazine shows like Inside Edition, Extra (TV program) and Dateline NBC, large conventions such as the Democratic National Convention, Republican National Convention or San Diego Comic-Con International, celebrity red-carpet events and sporting events.

EFP ranges from a camera operator or crew of two (camera operator with sound mixer) capturing high-quality imagery, to a multiple-camera setup utilizing videography, photography, advanced graphics and sound.

Sports

Sports television is one facet of EFP. Major television networks once owned their own production trucks for covering major events, but today, with the explosion in networks on cable and over-the-air, they rent television production trucks by the day or week from broadcast rental companies for more routine or remote broadcast productions.

DP Mark Schulze videotapes golf pro Phil Mickelson at Callaway hand-held with a Sony Betacam SP camera, 2005

A typical sports production truck includes:

- A large video switcher with an external digital video effects (DVE) unit and several mix/effect busses, to allow the Television director flexibility in calling for certain visual effects in the broadcast.
- Several tripod-mounted and handheld professional video cameras.
- A variety of zoom lenses for the tripod-mounted "hard" cameras, typically at least 50× to 100× magnification, and a maximum focal length of at least 600mm. The extreme amount of magnification is necessary because the cameras can be located quite a distance from the action.
- Several video recording and playback devices such as VCRs, hard disk recorders and video servers. Certain cameras or video feeds can be "isolated" to specific decks, and when something happens that the producer or director wants to see again, the deck can be rewound and shown on the air as an instant replay. Hard disk recorders typically allow some limited editing capabilities, allowing highlight reels to be edited together in the middle of a game.
- Several character generators allowing scores and statistics to be shown on screen. The scoreboards used in most sports facilities can be linked to the truck to drive the television production's graphics as well as the arena scoreboards.

- An audio mixing console booth and a variety of microphones to capture audio from the sportscasters and from the field of play.
- Several miles of various types of cable.

Television News Magazines

Television News magazines are longer and more in-depth TV programming than shorter "breaking news" clips that focus on an issue in a documentary style. They are driven by interviews of people who are directly involved in the topic covered and last for from 30 minutes to three hours.

Mark Wahlberg and Will Ferrell in the Press Room, San Diego Comic Con 2010

The first known television news magazine was Panorama on the BBC in 1953. Since then, the genre rose in popularity through the years including CBS's 60 Minutes, debuting in 1968. Its spin-off, 60 Minutes II debuted in 1999.

Electronic Field Production for a typical news magazine story may include one or several interviews with B-roll gathered typically by a three-person crew (producer, camera operator and audio technician/boompole operator). Locations vary. Typically the crew shoots the interview at the home or workplace of the interviewee. They may also go to additional locations that are a backdrop to the story. Lighting and shooting style are consistent with each's show's "look" or criteria.

Electronic Field Production shoot for Extra with DP Mark Schulze videotaping iScott Eastwood and Renee Bargh

After the interviews and B-Roll have been gathered, the producer may either hand-deliver the media to the studio, ship it by messenger service or a shipping company, or "feed it" in real time via a local satellite service.

Related Techniques

- Contrasted with the production values of EFP, in electronic journalism or electronic news-gathering (ENG), the emphasis is on quickness and agility in acquisition and rapidity in the process of editing, leading to final transmission to the audience is the goal. The two terms are often seen paired as EFP-ENG and vice versa.
- Many episodic television shows, four-camera situation comedy, television drama, such as PBS' Masterpiece Theatre all draw upon forms of EFP.

Electronic News-gathering

The TV4 and BBC HD satellite uplink DSNG at the IAAF World Athletics Championships in Helsinki, Finland

Microwave trucks seen transmitting. Modern news employs these trucks extensively.

On location outside Baltimore, cameraman Jim Furrer, sound recordist Bill Porter, and director David Ryan interview participants in a public rowing clinic as part of an early electronic journalism shoot in the 1980s.

Electronic news-gathering (ENG) is a broadcast news industry description of television producers, reporters and editors making use of electronic video and audio technologies for gathering and presenting news. The term was commonly used in the television news industry in the 1980s and 1990s, but it has since been less frequently used as the technology has become ubiquitous.

Electronic news-gathering can involve anything from a lone reporter taking a single professional video camera out to shoot a story, to an entire television crew taking a production or satellite truck on-location to conduct a live news report for an outside newscast.

The vehicle on which the electronic equipment is fitted is called DSNG (Digital Satellite News Gathering).

Beginnings

Shortcomings of Film

The term ENG was created as television news departments moved from film-based news-gathering to electronic field production technology in the 1970s. Since film requires chemical processing before it can be viewed and edited, it generally took at least an hour from the time the film arrived back at the television station or network news department until it was ready to be broadcast. Editing was done by hand on what was known as "color reversal" film, usually Kodak Ektachrome, meaning there were no negatives. Color reversal film had replaced black-and-white film as television itself evolved from black-and-white to color broadcasting. Filmo cameras were most commonly used for silent filming, while Auricon cameras were used for filming with synchronized sound. Since editing required cutting the film into segments and then splicing them together, a common problem was film breaking during the newscast. News stories were often transferred to bulky 2-inch videotape for distribution and playback, which made the content cumbersome to access.

Film remained important in daily news operations until the late 1960s, when news outlets adopted portable professional video cameras, portable recorders, wireless microphones and joined those with various microwave- and satellite truck-linked delivery systems. By the mid-1980s, film had all but disappeared from use in television journalism.

Transition to ENG

One of next-generation ENG mini-truck fleet of ABS-CBN.

Since ENG reduces the delay between the capturing of the footage and its subsequent broadcast, this allowed the news gathering and the reporting process to become one continuous cycle, with little pause between arriving at the site of a news story and putting the story in question on the air. Coupled with live microwave and/or satellite trucks, reporters were able to show live what was happening, bringing the audience into news events as they happened.

CNN launched in October 1980, as ENG technologies were emerging. The technology was still in its developmental stages, and had yet to be integrated with satellites and microwave relays, which caused some problems with the network's early transmissions. However, ENG proved to be a crucial development for all television news as news content recorded using videocassette recorders was easier to edit, duplicate and distribute. Over time, as editing technology has become simpler and more accessible, video production processes have largely passed from broadcast engineers to producers and writers, making the process quicker.

Television broadcast crews line up on a San Diego dock awaiting arrival of USS Vandegrift (FFG-48) crew after the sea rescue of the Kaufman family, 2014

However, initially the ENG cameras and recorders were heavier and bulkier than their film equivalents. This restricted the ability of camera operators from escaping danger or hurrying toward a news event. Editing equipment was expensive and each scene had to be searched out on the master recording.

Technology Developments

Using technology such as multicast or RTP over UDP, these systems achieve similar performance to high end-microwave. Since the video stream is already encoded for IP, the video can be used for traditional television broadcast or Internet distribution without modification (live to air).

As mobile broadband has developed, broadcast devices using this technology have appeared. These devices are often more compact than previous technology and can aggregate multiple mobile data lines to deliver a high definition-quality content live.

Broadcast Video Equipment

Easyrig Inventor Johan Hellsten with DP Mark Schulze wearing Easyrig at NAB 2015

The ongoing Technological evolution of broadcast video production equipment can be observed annually at the NAB Show in Las Vegas where equipment manufacturers gather to display their wares to people within the video production industry. The trend is toward lighter-weight equipment that can deliver more resolution at higher speeds. There has been an evolution from Film to Standard-definition television, High-definition television and now 4K.

Video production on Carlsbad beach using Sony FS7 camera, Sachtler carbon fiber tripod and Convergent Design Odyssey 7Q+ monitor

As of 2016, highlights included Unmanned aerial vehicles aka drones for the delivery of aerial footage, various lines of cameras that can deliver 4K resolution, graphics packages for news stations, which can be utilized inside their microwave and/or satellite vehicles, Wireless technology, POV cameras and peripherals from GoPro and other Action camera manufacturers, and the Odyssey 7Q+ monitor with Apollo multi-camera switcher/recorder from Convergent Design. This monitor fits on the back of a broadcast video camera and allows photojournalists to live-switch a multi-camera production in studio or on location.

Outside Broadcasts

Outside broadcasts (also known as "remote broadcasts" and "field operations") are when the editing and transmission of the news story are done outside of the station's headquarters. Use of ENG has made possible the greater use of outside broadcasts.

"Some stations have always required reporters to shoot their own stories, interviews and even standup reports and then bring that material back to the station where the video is edited for that evening's newscast. At some of these stations, the reporters sometimes even anchor the news and introduce the packages they have shot and edited."

News Reporters Gather at San Diego Home of James Holmes's Parents after Colorado theater shooting in Aurora, Colorado, July 2012

Short-form news stories are what local news reporters deliver to their stations. Longer-form stories about the same topics are covered by national or international broadcast news magazines such as Dateline NBC, 20/20, Nightline, 48 Hours, 60 Minutes and Inside Edition. Depending upon the scope of the story, the number of crews vying for position at the story venue (press conference, court house, crime location, etc.) can potentially be dozens.

Natural disasters, terrorism, death and murder are topics that reside at the top of the news-gathering hierarchy. For instance, in the U.S., the series of events culminating in what would thereafter be known as 9/11 galvanized every news division of every network.

"Especially on that first day, you were really just going to whomever had a piece of information," says "48 Hours" executive producer Susan Zirinsky, whose team produced the primetime coverage that first night. "You were getting cameras up, you were putting people in place, you were trying to wrap your brain around it. You wanted to step back and synthesize some of the information, which is what we were trying to do ... At that point, we thought there were many more dead, and it was still a search-and-rescue mission. It was a very, very complicated day to try to give context to."

"There is a hierarchy of news. It's a hierarchy of judgment, I guess. All deaths are equal to the victims and their families. But all deaths are not equal in the calculation of news value."

After she was lost at sea near San Clemente Island for 5 weeks, Luna the Dog is Reunited with Nick Haworth at San Diego Tuna Fleet Docks, March 2016

"Feel-good stories" such as the homecoming of a military parent, saving of a life, or selfless act of kindness, sometimes make their way into the news stream.

Microwave Spectrum Channels

In the United States, there are ten ENG video channels set aside in each area for terrestrial microwave communications. Use of these channels is restricted by federal regulations to those holding broadcast licenses in the given market. Channels 1 through 7 are in the 2 GHz band and channels 8, 9 and 10 are in the 2½ GHz band. In Atlanta for example, there are two channels each for the four news-producing television stations (WSB-TV, WAGA-TV, WXIA-TV, WGCL-TV), one for CNN, and another open for other users on request, such as Georgia Public Broadcasting.

Traditionally, the Federal Communications Commission has assigned microwave spectrum based on historic patterns of need and through the application/request process. With the other uses of radio spectrum growing in the 1990s, the FCC made available some bands of spectrum as unlicensed channels. This included spectrum for cordless phones and Wi-Fi. As a result, some of these channels have been used for news gathering by websites and more informal news outlets. One major disadvantage of unlicensed use is that there is no frequency coordination, which can result in interference or blocking of signals.

Audio Journalism

A common set-up for journalists is a battery operated cassette recorder with a dynamic microphone and optional telephone interface. With this set-up, the reporter can record interviews and natural sound and then transmit these over the phone line to the studio or for live broadcast.

Electronic formats used by journalists have included DAT, MiniDisc, CD and DVD. Minidisc has digital indexing and is re-recordable, reusable medium; while DAT has SMPTE timecode and other synchronization features. In recent years, more and more journalists have used smartphones or iPod-like devices for recording short interviews. The other alternative is using small field recorders with two condenser microphones.

Broadcast-safe

Broadcast-safe video (broadcast legal or legal signal) is a term used in the broadcast industry to define video and audio compliant with the technical or regulatory broadcast requirements of the target area or region the feed might be broadcasting to. In the United States, the Federal Communications Commission (FCC) is the regulatory authority; in most of Europe, standards are set by the European Broadcasting Union (EBU).

Broadcast Safe Video

Broadcast Safe Standard Definition Video

Broadcast Safe 625 Video

Broadcast safe standards for 625 lines of Standard Definition (Inaccurately referred to as PAL, a colour encoding that is usually used with such systems) video are:

- Common name = 625/50i (576i)
- Commonly used digital SD SDI baseband signal = SMPTE 259M-C, 270 Mbit/s bitrate
- Commonly used No. of Vertical Lines = 625 (576 visible active video)
- Commonly used Frame rate = 25 Hz (25 interlaced frame/s)
- Commonly used TV Resolution = 720 x 576 (576i)
- Black levels = 0 mV or 0 IRE

- White levels (Chrominance amplitude):
    - 700 mV p-p or 100 IRE - 100% intensity setting which corresponds to 100.0.100.0 color bars.
    - 75% intensity corresponding to 100.0.75.0 color bars, also referred to as EBU Bars.

Variants

Resolution		Aspect ratio	Pixel aspect ratio	Form of scanning	Framerate (Hz)
Vertical	Horizontal				
576	352	4:3 or 16:9	non-square	interlaced	25 (50 fields/s)
				progressive	25
	480	4:3 or 16:9	non-square	interlaced	25 (50 fields/s)
				progressive	25
	544	4:3 or 16:9	non-square	interlaced	25 (50 fields/s)
				progressive	25
	720	4:3 or 16:9	non-square	interlaced	25 (50 fields/s)
				progressive	25 50

Broadcast Safe 525 Video

Broadcast safe standards for 525 lines of Standard Definition (System M, NTSC) video are:

- Common name = 525/60i (480i)
- Commonly used digital SD SDI baseband signal = SMPTE 259M-C, 270 Mbit/s bitrate
- Commonly used Frame rate = 30 frame/s black and white, 29.97 interlaced frame/s color
- Black level = 7.5 IRE for NTSC in the US, 0 IRE in Japan.
- Blanking level = 0 IRE
- White levels = 100 IRE, 714 mV
- Maximum signal level = 120 IRE
- Minimum signal level = -20 IRE

Variants

Resolution		Aspect ratio	Pixel aspect ratio	Form of scanning	Framerate (Hz)
Vertical	Horizontal				
480	640	4:3	square	interlaced	29.97 (59.94 fields/s) 30 (60 fields/s)
				progressive	23.976 24 29.97 30 59.94 60
	720	4:3 or 16:9	non-square	interlaced	29.97 (59.94 fields/s) 30 (60 fields/s)
				progressive	23.976 24 29.97 30 59.94 60

Broadcast safe High Definition Video

Digital broadcasting is very different from analog. The NTSC and PAL standards describe both transmission of the signal and how the electrical signal is converted into an image. In digital, there is a separation between the subject of how data is to be transmitted from tower to TV, and the subject of what content that data might contain. While data transmission is likely to be a fixed and consistent affair, the content could vary from High Definition video one hour, to SD multicasting the next, and even to non-video datacasting. In the US, 8VSB transmits the data, while MPEG-2 encodes pictures and sound.

Resolution		Aspect ratio	Pixel aspect ratio	Form of scanning	Framerate (Hz)
Vertical	Horizontal				
720	1280	16:9	square	progressive	23.976 24 25 29.97 30 50 59.94 60
1080	1920	16:9	square	interlaced	25 (50 fields/s) 29.97 (59.94 fields/s) 30 (60 fields/s)
				progressive	23.976 24 25 29.97 30

Broadcast Safe Audio

Broadcast engineers in North America usually line up their audio gear to nominal reference level of 0 dB on a VU meter aligned to +4dBu or -20dBFs, in Europe equating to roughly +4 dBm or -18 dBFS. Peak signal levels must not exceed the nominal level by more than +10 dB.

Broadcast audio as a rule must be as free as possible of Gaussian noise, that is to say, it must be as close to the noise floor, as is reasonably possible, considering the storage or transmission medium.

Broadcast audio must have a good signal-to-noise ratio, where speech or music is a bare minimum of 16db above the noise of the recording or transmission system. For audio that has a much poorer signal-to-noise ratio (like cockpit voice recorders), sonic enhancement is recommended.

Non-standard Video

Although almost any video gear can create problems when broadcast, equipment aimed at consumers sometimes produces video signals which are not broadcast safe. Usually this is to reduce the cost of the gear, since a non-standard video signal in the home might not create the problems that one might find in a broadcast facility. Potential flaws exist with:

- VHS and 8 mm: Consumer devices generally lack time base correction that may cause problems with genlock and sync with some analogue and digital broadcast equipment. Consumer analogue video systems have greater system noise and lower chrominance and luminance than is normal for standard definition TV. As a general broadcast engineering rule all analogue videotape origin material should be genlocked before transmission, but this is not mandatory or necessary for all conditions. All analogue videotape by default is broadcast safe under normal playing conditions.
- Older videogame systems: Video game consoles before the sixth generation and 8-bit home computers generated a video signal lacking the half scan line needed to make interlace happen. This subtle simplification caused NTSC sets to scan 240p/60 instead of 480i60, with similar results for PAL. While this actually improved picture quality for the kind of low-definition images that videogames of this era generated, such a signal modification could cause problems in a broadcast environment as the signal behaviour is outside the original television system specifications. Genlocking—but not timebase correction—are the recommended broadcast engineering solutions.
- Computer video signals: Computer video can be set up to run at many different frame or field rates, ranging 50 frame/s to more than 240 frame/s. Computer video is generally progressive by default, but many interlaced modes exist. A scan converter is typically needed to convert these signals to one of the many acceptable broadcast standards, such as 59.97 Hz or 50 Hz. This type of conversion typically degrades the quality of the broadcast image, usually resulting with either "motion artifacts" or a lower resolution. It is recommended that the display rate be set to equal the target television rate if possible.

In Digital Television Only Environments

In nations that have fully converted to digital television, broadcast safe analogue television takes on a slightly different meaning. All broadcasting systems will have been mostly converted to digital only outputs, leaving fewer entry points for analogue television signals.

What this means is that all devices that feed to the television transmitter must take in and feed standard analogue television signals into the transmission chain. Mostly it is up to the switcher to notify if there is non-broadcast safe video to the programmer. However, due to the limitations of many switchers for DTV and HDTV it ultimately is up to the automation systems to alert the programmer of non-broadcast safe video inputs.

As a matter of broadcast engineering practice, 4:3 analogue television signals will always pose the most problems with broadcast safe compliance. The use of portable and cheap timebase-genlock systems for analogue television inputs in the digital television studio will be clearly mandatory for the next 50 years.

Broadcast Delay

Many US radio talk shows use broadcast delay to avoid FCC penalties

In radio and television, broadcast delay refers to the practice of intentionally delaying broadcast of live material.

Usage

A short delay is often used to prevent profanity, bloopers, violence, or other undesirable material from making it to air, including more mundane problems such as technical malfunctions (i.e. an anchor's lapel microphone goes dead) or coughing. In this instance, it is often referred to as a seven-second delay or profanity delay.

Longer delays can also be introduced, often to allow a show to air at the same time for the local market as is sometimes done with nationally broadcast programs in countries with multiple time zones. That can sometimes be simply achieved with a video tape recorder or similar technology. In the context of modern digital video recorders, this can now be considered a class of time shifting. In this instance, it is often called a tape delay or west-coast delay in the United States, so-called because special events (including award shows) broadcast live in the Eastern or Central time zones of the US are often tape-delayed on the west coast. This includes Southern California (with the exception of the Academy Awards, which airs live in Los Angeles and the rest of the west coast), despite the fact that Southern California is where many live televised events (*American Idol, Dancing With The Stars*) take place. Nationally telecast morning news shows in the US (such as *CBS This Morning*, *Today*, and *Good Morning America*) typically are aired live only in the Eastern time zone, while on tape delay in the remaining time zones. This allows post-production staff to edit out any glitches that occurred during the live broadcast.

Tape delay also refers to the process of broadcasting an event at a later scheduled time. This is because either a scheduling conflict prevents a live telecast, or a broadcaster seeks to maximize ratings by airing an event in a certain timeslot. This can also be done due to time constraints where certain portions (usually those that do not affect the outcome of the show) are edited out, or availability of hosts or other key production staff only at certain times of the day (e.g. *O'Reilly Factor* except during special events). For example, during the Sydney Olympics, daytime events were occurring at early morning hours in the Americas, Africa, and Europe, so some broadcasters showed high-profile events twice (live, then rebroadcast during prime time), while others withheld the same event to be broadcast solely during prime time. Sporting events aired on tape delay are often edited down for time considerations, highlighting what the broadcaster feels are the most interesting portions of the event, or advertising.

History

The radio station WKAP in Allentown, Pennsylvania, introduced a tape delay system consisting of an external playback head, which was spaced far enough away from the record head to allow for a six-second delay. A system of rollers guided the tape over the playback head before it wound up on the take up reel. This system was introduced in 1952 when WKAP started a talk show called *Open Mic*. It is believed that this was the first time a telephone call-in show was broadcast with the telephone conversation "live" on the air. The FCC rules at the time prohibited the broadcasting of a live phone conversation. However, there was no rule prohibiting a taped playback of a phone call, provided that a "beep" tone was heard by the caller every 15 seconds so that the caller knew he was being recorded. The six-second delay constituted a "taped" phone conversation, thus complying with FCC regulations, that being a legal fiction.

The broadcast profanity delay was invented by C. Frank Cordaro (July 13, 1919 - February 20, 1997) who was Chief Engineer of WKAP, originally on AM 1320, in Allentown, Pennsylvania, during the 1950s and early 1960s. Ogden Davies, then General Manager of WKAP, assigned Cordaro the task to develop a device whereby profanity during a "live" conversation could be deleted by the radio talk show host before it was broadcast. This new device was to be used on the *Open Mic* radio talk show. The device Cordaro developed was the first tape delay system. WKAP was one of several stations owned by the Rahal brothers of West Virginia (later Rahal Communications). First tested and used at WKAP, this tape system for broadcast profanity delay was then installed at the other Rahal-owned radio stations. From the Rahal brothers' stations, the broadcast profanity delay went into common usage throughout the US.

John Nebel, who began a pioneering radio talk show in New York City in 1954, was one of the early users of a tape delay system.

Computerized Delay

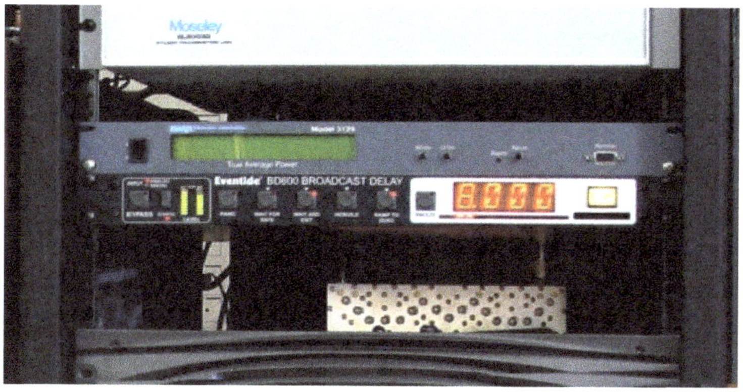

Eventide BD600 Broadcast Delay

In 1977, the capacity of RAM (random-access memory) had reached 16 kb (kilobits) per chip, enough to think about using computerized digital audio means to create a sufficient delay for content deletion. By storing audio digitally, it was possible to move a "virtual tape head" along recorded audio. Eventide, Inc. created the first digital broadcast delay for this purpose. The device (known colloquially as a "dump box") had a large "DUMP"/"DELAY DUMP" button that would bring the delay to zero, thus removing unwanted segments. In addition to this convenience, it would also "rebuild" the delay time by unnoticeably lengthening the normal pauses in spoken material. Thus, a minute or so later, the broadcaster would again have full delay, often leaving the listener unaware that material had been deleted.

In modern systems, a profanity delay can be a software module manually operated by a broadcast technician that puts a short delay (usually thirty seconds) into the broadcast of live content. This gives the broadcaster time to censor the audio (and video) feed. This can be accomplished by cutting directly to a non-delayed feed, essentially jumping past the undesired moment (something that can be quite jarring to a viewer or listener). In other cases, dedicated hardware units similar to the original digital unit but with improved quality and editing capability can be used. These products can even "build up" delay with difficult program material such as music. Alternatively, a bleep noise or other substitute sound can be inserted. This is more difficult to do with live content, however, and more often appears on recorded material.

Bleep Censor

A bleep censor is the replacement of a profanity or classified information with a beep sound (usually a 🔊 1000 Hz tone (help·info)) in television and radio.

History

Bleeping has been used for many years as a means of censoring TV programs to remove content not deemed suitable for "family" or "nighttime" viewing and personal information for privacy. The bleep censor is a software module, manually operated by a broadcast technician. A bleep is sometimes accompanied by a digital blur or box over the speaker's mouth in cases where the removed speech may still be easily understood by lip reading.

Censor boxes, such as the one above, may be used along with the bleeps so that the audience would not lip-read the swearer's words. Above, the cartoon says "Oh-", followed by the censor.

On closed caption subtitling, bleeped words are usually represented by the phrase "(*bleep*)", sometimes the phrase "[expletive]" or "[censored]", occasionally hyphens (e.g. *f—k f*---), and sometimes asterisks (e.g. ****, f***, f**k, f*ck, f#@k or f#@%), remaining faithful to the audio track. Where open captions are used (generally in instances where the speaker is not easily understood), a blank is used where the word is bleeped. Occasionally, bleeping is not reflected in the captions, allowing the unedited dialogue to be seen. Sometimes, a "black bar" can be seen for closed caption bleep, or pixelization.

Bleeping is normally only used in unscripted programs – documentaries, radio features, panel games etc. – since scripted Drama and comedy are designed to suit the time of broadcast. In the case of comedies, most bleeping may be for humorous purposes.

When films are edited for daytime TV, broadcasters usually prefer not to bleep swearing, but cut out the segment containing it, replace the speech with different words, or cover it with silence or a sound effect. In the first example, the film may (unintentionally) become nonsensical or confusing if the removed portion contains an element important to the plot. The bleep is sometimes used for privacy reasons, concealing things such as names and addresses. Bleep censors have seen a rare use in cinema film, for example during Johanna Mason's interview in *The Hunger Games: Catching Fire*, or during Tony Stark's meeting with Senator Stern in *Iron Man 2*.

Bleeping is commonly used in English-language and Japanese-language broadcasting, but is sometimes/rarely used in some other languages (such as Dutch, Filipino, French, German, Italian, Portuguese, Spanish, Russian and Polish), displaying the varying attitudes between countries; some are more liberal towards swearing, less inclined to use strong profanities in front of a camera in the first place, or unwilling to censor. In the Philippines and Ecuador, undubbed movies on television have profanity muted instead of bleeped.

Regulations

Advertising in the United Kingdom

Television and radio commercials are not allowed to use bleeps to obscure swearing under BACC/CAP guidelines. However, this does not apply to program trailers or cinema advertisements and "fuck" is beeped out of two cinema advertisements for Johnny Vaughan's Capital FM show and the cinema advertisement for *Family Guy* season 5 DVD. An advert for Esure insurance released in October 2007 uses the censor bleep, as well as a black star placed over the speaker's mouth, to conceal the name of a competitor company the speaker said she used to use. The Comedy Central advert for *South Park: Bigger, Longer & Uncut* had a version of "Kyle's Mom's a Bitch" where vulgarities were bleeped out, though the movie itself did not have censorship, and was given a 15 rating, despite a high amount of foul language.

A Barnardo's ad, released in summer 2007, has two versions: one where a boy can be heard saying "fuck off" four times which is restricted to "18" rated cinema screenings, and one where a censor bleep sound obscures the profanity which is still restricted to "15" and "18" rated films. Neither is permitted on UK television.

Trailers for programs containing swearing are usually bleeped until well after the watershed, and it is very rare for any trailer to use the most severe swearwords uncensored.

The UK version of the Adventureland Red Band trailer (the version shown in cinemas) which showed before *Funny People* and *Drag Me to Hell* when it was out in UK cinemas had the profanities bleeped out in order to have a 15 certificate.

United States

The Federal Communications Commission has the right to regulate indecent broadcasts. However, the FCC does not actively monitor television broadcasts for indecency violations, nor does it keep a record of television broadcasts. Reports must be documented exclusively by the public and submitted in written form, whether by traditional letter or e-mail.

The FCC is allowed to enforce indecency laws during 6 a.m. – 10 p.m. local time. In addition, for network broadcasts, offensive material seen during watershed in one time zone may be subject to fines and prosecution for stations in earlier time zones: for instance, a program with offensive content broadcast at 10 p.m. Eastern Time/Mountain may fall out of watershed at 9 p.m. Central Time/Pacific Time. To compensate, a channel may only air uncensored material after 1 a.m. Eastern Time so that the broadcast is out of watershed in the contiguous United States.

Cable and satellite channels are subject to regulations on what the FCC considers "obscenity," but are exempt from the FCC's "indecency" and "profanity" regulations, though many police themselves, mainly to appeal to advertisers who would be adverse to placing their ads on their programs.

New Zealand

The Office of Film and Literature Classification enforces what can and cannot be said on television in New Zealand and order television networks to apply the bleep censor over objectionable material.

References

- San Diego Union Tribune, "Lost at sea and presumed dead, rescued dog is returned to San Diego," by Debbi Baker, March 16, 2016.
- Elly, Wally (2006-09-20). "In local radio, change is common - with one exception". The Morning Call. Retrieved 2016-08-16.

- Bill Keller, "Is There A Hierarchy of the Importance of Death in the News Business?" by David Folkenflik, NPR, November 17, 2015.
- "bleep – definition of bleep by the Free Online Dictionary, Thesaurus and Encyclopedia". Thefreedictionary.com. Retrieved 2012-01-19.

Permissions

All chapters in this book are published with permission under the Creative Commons Attribution Share Alike License or equivalent. Every chapter published in this book has been scrutinized by our experts. Their significance has been extensively debated. The topics covered herein carry significant information for a comprehensive understanding. They may even be implemented as practical applications or may be referred to as a beginning point for further studies.

We would like to thank the editorial team for lending their expertise to make the book truly unique. They have played a crucial role in the development of this book. Without their invaluable contributions this book wouldn't have been possible. They have made vital efforts to compile up to date information on the varied aspects of this subject to make this book a valuable addition to the collection of many professionals and students.

This book was conceptualized with the vision of imparting up-to-date and integrated information in this field. To ensure the same, a matchless editorial board was set up. Every individual on the board went through rigorous rounds of assessment to prove their worth. After which they invested a large part of their time researching and compiling the most relevant data for our readers.

The editorial board has been involved in producing this book since its inception. They have spent rigorous hours researching and exploring the diverse topics which have resulted in the successful publishing of this book. They have passed on their knowledge of decades through this book. To expedite this challenging task, the publisher supported the team at every step. A small team of assistant editors was also appointed to further simplify the editing procedure and attain best results for the readers.

Apart from the editorial board, the designing team has also invested a significant amount of their time in understanding the subject and creating the most relevant covers. They scrutinized every image to scout for the most suitable representation of the subject and create an appropriate cover for the book.

The publishing team has been an ardent support to the editorial, designing and production team. Their endless efforts to recruit the best for this project, has resulted in the accomplishment of this book. They are a veteran in the field of academics and their pool of knowledge is as vast as their experience in printing. Their expertise and guidance has proved useful at every step. Their uncompromising quality standards have made this book an exceptional effort. Their encouragement from time to time has been an inspiration for everyone.

The publisher and the editorial board hope that this book will prove to be a valuable piece of knowledge for students, practitioners and scholars across the globe.

Index

A

Advertising, 3, 19-21, 23, 30, 45, 57-62, 64, 66, 68-73, 75-76, 78-79, 96, 98, 109, 113-114, 119, 123, 137, 140, 144, 146, 153, 162, 165-204, 226, 228

Apple Trademark Protections, 160

Apps, 151-152, 154, 157

Audio Codec, 45-48, 50, 52, 55, 162

Audio Quality, 45, 48, 52-55, 137, 162

B

Bands, 16, 21, 32, 34, 36-37, 41, 43, 46-47, 78, 96, 155, 178, 221

Bit Rates, 46, 49, 51, 54

Bleep Censor, 205, 227-229

Block Programming, 114

Blogs, 132, 144-145, 147-148, 162, 197

Bridging, 114

Broadcast Automation, 10-11, 113, 205, 207, 210

Broadcast Journalism, 132-133, 139-140

Broadcast Network, 7, 10, 97, 123, 205, 207

Broadcast Signal Intrusion, 38, 125, 128

Broadcast Towers, 12

Broadcast-safe, 205, 221

Broadcasting, 1, 3-10, 12-13, 16-30, 32-45, 51-52, 54-55, 57-79, 82-83, 86, 89-92, 95-98, 108-110, 121-124, 127-131, 140, 156, 168-169, 171-175, 177-178, 182-183, 185, 205-207, 209-211, 217, 220-221, 223, 225-226, 228

C

Cellsodes, 155

Commercial Broadcasting, 6, 38, 40, 57-59, 61, 109, 168, 182

Compatibility, 56, 151, 201

Consoles, 12, 224

Control Rooms, 12

Copyright, 70, 118, 152

Counterprogramming, 107, 114, 131

Coverage, 21, 47, 49, 52, 55-56, 60, 68, 70, 74, 123-125, 132, 137, 146, 200, 220

Crossprogramming, 114

D

Dayparting, 114

Digital Engineering, 12

Digital Journalism, 132, 142-146, 148, 163-164

Digital Marketing, 189

Display Ads, 191-193, 196, 200

Display Advertising, 190-191, 193, 203

E

Electronic Field Production, 95, 141, 205, 211, 213-215, 217

Electronic Media, 1-3, 5, 10, 59, 68, 132, 153, 165

Electronic Publishing, 132, 150-152

Enhanced Podcasts, 160

Error-correction Coding, 47-48

Expanding Ad, 192

G

Games, 3, 24, 31, 100-101, 122-124, 153-154, 190, 196, 200, 204, 228

Global Commercial Broadcasting, 58

H

Hammocking, 115

History of Broadcasting, 16, 18, 23, 27, 206

Hotswitching, 115

I

Images, 50, 134, 154, 160, 187, 200, 224

Impact on Publishers, 145

Impact on Readers, 144

Influence on Television Schedules, 106

Ip Issues in Trademark, 159

J

Journalism, 3, 8-9, 132-134, 136-137, 139-140, 142-149, 162-164, 216-217, 221

L

Live Forms, 6

Local Television, 112, 118, 141

M

Medium Wave, 16, 19, 21-23, 35, 41-42, 96

Microbroadcasting, 38, 96

Microwave Spectrum Channels, 220

Mobile Advertising, 186-188, 190-191, 195-196, 199, 203

Mobile Content, 132, 153-154, 156-157

Mobile Rich Media, 187
Modes, 18, 47, 56, 201, 224
Modulation, 42, 46-47, 49
Monetary Rates, 108
Music, 3, 16, 18-23, 25-30, 32, 36, 43, 53-54, 57-58, 65, 74, 77-78, 84, 100, 109-111, 119, 121, 123, 129, 140-141, 151, 153-157, 160, 169, 178-179, 182-183, 208-210, 224, 227

N
News Feed Ads, 193
News Jobs, 141
Non-standard Video, 224

O
Online Advertising, 144, 153, 190-191, 193-194, 196, 198-200, 202-204
Outside Broadcasting, 205, 211

P
Personal Audio Lawsuits, 160
Podcast, 152, 157-163
Podcast Novels, 160-161
Power Requirements, 56
Protocol Stack, 47

R
Radio Advertisement, 182
Radio Broadcasting, 4-5, 7-8, 23-25, 28, 30, 38, 40, 44, 51, 57, 67-68, 72, 74, 96, 210
Radio Syndication, 109
Ratings, 58-59, 79, 98, 104-108, 112-113, 116-117, 123, 145, 167, 183-184, 207, 226
Reception Quality, 45-46, 48-49, 54-55
Recorded, 1, 6-8, 45, 110, 117, 122, 140-141, 155, 161-162, 178-180, 185, 198, 205-206, 212, 217, 226-227
Remote Broadcast, 11, 38, 95, 141, 211, 214
Rerun, 97, 117-120, 168

S
Search Ads, 191, 196
Signal Delay, 55
Simulcast, 5, 7, 38, 103, 121-125, 209
Single-frequency Networks, 49
Social Impact, 7
Sound Quality, 45, 52-53
Spoiling, 115
Stacking, 115
Streaming, 2, 4, 7, 38, 70, 74, 131, 156-157, 180, 185, 191, 205
Stripping, 115

T
Television Advertisement, 165-166, 169-170, 177, 179
Television Schedules, 106
Television Signal Intrusions, 129
Tentpoling, 116
Theming, 116
Transmission Costs, 55
Trick Banners, 193
Tv Listings, 113, 120

V
Variable Bandwidth, 54
Video, 1-3, 5-6, 10-12, 16, 50-51, 70, 95, 118-119, 121-127, 129-130, 132, 137, 139-145, 153-157, 161-163, 166, 179-181, 190, 192, 197, 199-201, 211-214, 217-227
Viral Marketing, 188